Game Playing
With
Computers

Revised Second Edition

DONALD D. SPENCER

President
Abacus Computer Corporation

HAYDEN BOOK COMPANY, INC.
Rochelle Park, New Jersey

To my wife RAE
and our six game players
SANDRA, SUSAN, SHERRIE, STEVEN,
LAURA, and MICHAEL

Library of Congress Cataloging in Publication Data

Spencer, Donald D
  Game playing with computers.

  Bibliography: p.
  1. Mathematical recreations—Data processing.
I. Title.
QA95.S62   1975   793.7'4   75-31506
ISBN 0-8104-5103-4

Printed in the United States of America

    1   2   3   4   5   6   7   8   9   PRINTING
  ──
    75 76 77 78 79 80 81 82 83   YEAR

PREFACE

It has been my conviction that most members of the computer programming community are also game players. Computerized game playing may be found to some degree at almost every computer installation. This is primarily because most computer professionals agree that information gained while programming computers to play games is directly transferable to other areas of scientific and business programming.

Game Playing With Computers is intended to introduce to the reader many games that may be programmed for a digital computer. Since game playing is an excellent media for one to learn computer programming, it may be used by students and beginning programmers. On the other hand, since many game-playing programs are extremely complex, the book may also be used by senior programming personnel, system analysts and mathematicians.

Chapter 1 introduces computerized game playing, with examples cited from popular game-playing computer programs. The next four chapters describe many game-playing programs: Magic Squares, Prime Numbers, 15 Puzzle, 34 Puzzle, checkerboard games, binary games, and so forth. These programs, written in FORTRAN and fully tested, are developed in a general manner so they may be studied by the reader. FORTRAN is used rather than any other computer programming language because of its increasingly widespread adoption by computer manufacturers and users.

The six chapters of Part 2 describe and illustrate approximately 50 games that may be programmed for computer solution. Among these game descriptions are Keno, Roulette, Slot Machines, Chuck-A-Luck, Baccarat, Faro, Bingo, Go-Moko, The King's Magic Tour, Blackjack, Perfect Numbers, Chinese Fan Tan, Crown and Anchor, Nim, Pentomino Game, Chess, Go, Tic-Tac-Toe, Checkers and many others.

Part 3 describes several game-playing topics such as betting systems and random number generation. Chapter 12 discusses game playing on a time-sharing computer system using the BASIC programming language. Chapter 15 includes several game-playing exercises.

Concluding the book are a bibliography and a section containing answers to the odd-numbered exercises.

Preface

Practically none of the games described in this book is original with me. I am indebted to all those listed in the bibliography who have developed and explored the games and mathematical recreations that are included in this book.

To my wife Rae I extend my deepest appreciation for her typing assistance in helping me complete this book. Also, I wish to thank my wife and children for their patience and understanding, without which this book would never have been written.

One word of advice for those readers who will use the games described in this book: be sure you fully understand a game before attempting to write a game-playing computer program. Always draw a flowchart of the game logic *prior* to coding the solution in a programming language. These simple steps will allow you to more fully enjoy solving games with a computer.

Donald D. Spencer

Ormond Beach, Florida

CONTENTS

PART ONE

Game-Playing Programs

CHAPTER 1

Game Playing With Computers

1.1 Introduction to Game Playing

THE ORIGIN OF GAMES has been vaguely assigned to the inborn tendency of mankind to amuse itself. Games have no geographical boundaries and game playing is found in all parts of the world whether it be in the underdeveloped areas of Africa or in a plush New York apartment penthouse.

Modern games have so nearly lost their original meaning that even in the light afforded by history it is practically impossible to trace their origin. Most of the games that are presently played in America have origins in China, Korea, Japan, Greece, Italy and Africa.

Up until the invention of the electronic digital computer, game playing was primarily restricted to mere humans or special purpose machines. Today, mathematicians, programmers, scientists and game-playing novices are spending a considerable amount of time programming general purpose digital computers to play games. Although game playing and the writing of game-playing computer programs is fun, there are two other reasons why humans are using digital computers to play games.

As stated earlier, *games are fun to play* and practically everyone has his own repertoire of games. First of all, the rules of how to play a game are usually relatively simple and can be understood in a minimum amount of time. A person learning to write programs for digital computers should begin with problems that are well defined and easily understood. Thus, game programs provide us with excellent problems for learning computer programming. The beginning programmer can

1

understand the problem to be programmed in a minimum amount of time; therefore, he can devote more time to the learning of the computer, the programming language and the techniques of problem solving with a computer. A game is a closed universe, with a set of explicitly stated rules and a fixed goal. Also, the means of winning the game or reaching the fixed goal during a particular play of the game are usually known.

The second reason why humans are using computers to play games is that games are often good analogies to actual situations involving humans and their environment. Gaming is being applied to business management and war strategy. Business executives are playing games with digital computers that simulate the operation of their business. Games of this type allow the executive to keep an active study of his employees, to learn more about his company and to simulate all activities of his company. Consider the problem of determining a plan to plant several crops and distribute the produce after it has been harvested. Without the use of a computer, the problem must be solved by a trial and error method and could take up to a year before one would know if the plan was a successful one. However, a game situation that simulated the actual situation on a small scale could be programmed for a computer and many different plans could be checked until a proper plan was determined.*

Information learned with programs such as *Chess, Checkers, Go* and *Go-Moko* may very well apply to a class of other problems. Consider the problem of the maintenance and repair of electronic equipment. Assuming that all parts of a chassis could be checked or monitored by automatic means, a computer could be programmed to do the automatic checking function. Using a learning program similar to a Chess, Go or Go-Moko program, the computer could indicate which parts are likely to go bad first for a specific chassis, and also which sections appear to be weakest. For example, in a certain run of radios a lot of weak diodes may have been used. The computer can be programmed, after a few of these are found to be bad, to look at this part first each time one of these radios comes up. After the run is over, the computer will return to its normal pattern. The computer might reveal that in audio amplifiers it is the output tubes that go first and hence immediately check the output tubes of all audio amplifiers of any design. These frequency-of-occurrence techniques along with standard techniques such as signal tracing and analysis of symptoms of malfunctionings could make a computer acting as a technician more efficient than a human technician.

War games have been used by military organizations for many years for training personnel and testing military plans. War gaming was used

* Small problems of this type may also be solved by hand computations.

by the Germans and Japanese prior to World War II. However, it was in America (about 1950) that computers were first used to develop simulations of military operations.

It is difficult to determine exactly how learning techniques of game-playing programs can be applied to actual problems. Perhaps these programs will solve many problems that humans have thus far been unable to solve. Such was the case when a computer was directed to execute a geometry theorem-proving program. This program found proofs for theorems in plane geometry. When the program was used to prove that the two base angles of an isosceles triangle are equal, it produced a proof which was very surprising to the programmer who developed the program. The proof it produced was shorter than the proof that is usually given in basic geometry textbooks. Of course, the programmer could have studied the program and determined exactly what proof would have been generated by the program. This may have taken years to do, however, as the program was designed to solve many different problems.

Games may be classified in a variety of ways. Dr. Eric Berne in *Games People Play*** describes games as events that people play in their everyday lives. Dr. Berne's type of games falls in the area of trans-actional game analysis. There is a branch of mathematics known as game analysis that deals with the mathematical theory of games. Games of strategy and logic are typical of games found in this area. An ele-mentary school teacher considers dodge-ball a game while the college coach would be insulted if everyone did not recognize football as a game. The gamblers in Las Vegas and Monte Carlo are commonly found indulging in casino games.

I have used *games* very broadly to classify a large collection of rec-reations. In this book games will include: *popular games* such as Chess, Go, Tic-Tac-Toe and Checkers; *casino games* such as Roulette, Craps, Blackjack and Keno; *mathematical recreations* such as Prime Numbers, Magic Squares, Pentominoes and The Knight's Tour; and *puzzles* such as the 15 Puzzle. One additional classification will be used in this book as all games will fall into two categories: (1) games for which there is a known algorithm, and (2) games for which there are no known algorithms, other than the simple *process of exhaustion.†* Nim, *Magic Squares, Tic-Tac-Toe, Roulette* and many *mathematical puzzles* fall into the first category. *Chess, Checkers* and the Japanese game of *Go* fall into the second category.

Before we proceed any farther, I will define an algorithm. *An algorithm is a list of instructions or rules that specifies a sequence of*

---

* All book references will be found in the Bibliography.

† A process where every possible game configuration is known and considered during game play.

operations which will give a correct answer to a problem of a given type. Games that have no known algorithm are often solved by heuristic processes, i.e., the computer uses a process which seems reasonable but is probably not always optimum. This process may solve a specified game and usually does but does not guarantee a solution the way an algorithm does. Heuristic programming is used in many areas other than game playing (pattern recognition, theorem proving) and is responsible for the development of list-processing programming languages, such as LISP and IPL.

The word *algorithm* is derived from the name of a ninth-century Arabian mathematician, al-Khowarizmi. He was interested in solving certain problems in arithmetic, and he devised a number of methods for solving them. These methods were presented as a list of specific instructions, and his name has become attached to such methods. al-Khowarizmi wrote the book *al jabr w' al muquabalah* (the reunion and the opposition). These words referred to the two main processes employed in solving "equation" problems, *reunion* being presumably the bringing together of terms involving the unknown quantity, *opposition* being the final stage, when a "reunited" unknown quantity was faced by some number. The book was translated into Latin, and the Latin title of the book was eventually reduced to our familiar word *algebra.*

A recipe in a cookbook is a good example of an algorithm. The preparation of a certain dish is broken down into many simple instructions or steps that anyone experienced in cooking can understand. Another example of an algorithm is the procedure used to play the following game. You and your opponent take alternate turns in removing matches from a pile. The minimum number of matches that can be removed from the pile is 2 and the maximum number is 6.* The winner of the game is the one who forces his opponent to take the *last* match. For example, consider the progress of the following game with an original pile of 76 matches:

You take 3 and leave 73
Opponent takes 5 and leaves 68
You take 3 and leave 65
Opponent takes 6 and leaves 59
You take 2 and leave 57
Opponent takes 6 and leaves 51
You take 2 and leave 49
Opponent takes 6 and leaves 43
You take 2 and leave 41
Opponent takes 4 and leaves 37
You take 4 and leave 33

---

* There are many variations of this game. Any number of matches may be in the original pile, and the number of matches that can be picked up may also vary. The game is a special case of Nim.

Opponent takes 6 and leaves 27
You take 2 and leave 25
Opponent takes 5 and leaves 20
You take 3 and leave 17
Opponent takes 4 and leaves 13
You take 4 and leave 9
Opponent takes 2 and leaves 7
You take 6 and leave 1
Opponent takes 1

Since the opponent moved last, you win. For this game there is a set of rules that will insure that you always win. Such a set of rules is called an algorithm.

We solve most decision problems by using algorithms, that is, by breaking the problem down into many unambiguous steps. Problems for computer solution likewise must be broken down into many simple steps or instructions. Although the algorithms for many problems are rather simple, the algorithms for complicated scientific problems can be quite complex. The algorithm for determining if 639 is a prime number is simple. One only has to divide 639 by each of the numbers 1, 2, 3, . . . , 638, 639.* If any division results in a zero remainder, the number is not prime; otherwise it is a prime number. The algorithm for playing chess, on the other hand, is complex and involves many steps.

Consider the problem of Königsberg Bridges. The map in Figure 1-1(a) shows the Prussian city of Königsberg and the river loop that divides it into four areas (marked A, B, C, and D). Connecting the areas are seven bridges. The dotted lines indicate all the possible routes between A, B, C, and D using the bridges. It has been a tradition among the townspeople that the seven bridges could not all be crossed in a continuous walk without recrossing the route at some point, but no one knew the explanation. When Leonhard Euler, a Swiss mathematician, heard of the Königsberg bridges, he realized that an important principle was involved, and he proceeded to design an algorithm which demonstrated why such a walk was impossible.

In order to understand the solution, consider the diagram in Figure 1-1(b). The lettered points and the lines of this diagram correspond to those marked on the map except that each line in Figure 1-1(b) represents a bridge. Each point, called a *vertex*, represents a land region. The diagram has four vertices and seven lines. Now suppose we can travel over each bridge once and only once. If the total route starts from A, we might go first to B. We would enter B on one line and leave on another line. Thus, each time we pass through a vertex, there must be exactly two lines connected to that vertex.

After the total path is drawn, we can say: Every vertex has an even

---

* There are much more efficient algorithms for this problem.

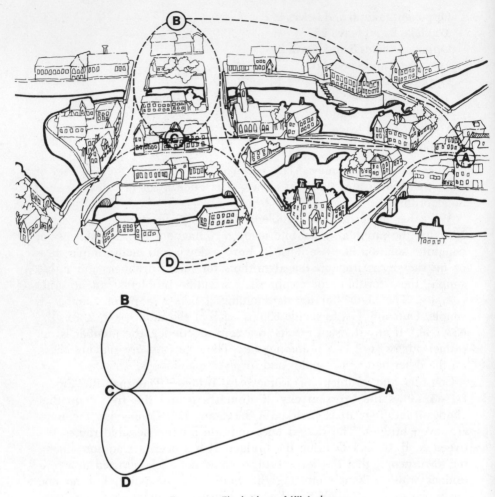

Figure 1-1. The bridges of Königsberg.

number of lines connected to it. The only possible exceptions are the vertex from which we start and vertex at which we end. Thus a closed path covering all bridges is possible only if:

1. Every vertex has an even number of lines (then we start and end at the same point), or

2. Exactly two vertices have an odd number of lines (then we start at one of these and end at the other).

Figure 1-1(b) reveals that the vertices have the following numbers of lines: A—3, B—3, C—5, and D—3. Since, all four numbers are odd, there is no hope of walking over all seven Königsberg bridges without retracing any path.

Although honest machines have been built to play games, game play-

Figure 1-2. Baron Wolfgang von Kempelen's chess playing machine, which never lost a game. The picture shows the system of levers used by the cramped operator to control the machine.

ing machines have had a long and shady history. From Baron Wolfgang von Kempelen's chess playing automaton at the beginning of the nineteenth century to the one-armed bandits of today, they have tended to honor the rules of the game more in the breach than the observance. Von Kempelen's machine (Figure 1-2) played good chess—it is said to have beaten Napoleon—but concealed inside was a legless Pole, who controlled the machine through ingenious mechanisms. Only in the last twenty years have machines really been developed which can play games according to the rules and win.

Charles Babbage, early in the nineteenth Century, attempting to find a way of financing his Analytical Engine, wanted to build a *Tic-Tac-Toe* machine and tour the country to exhibit it in towns for a fee. In 1914, a Spanish inventor named L. Torres y Quevedo constructed an honest *chess-playing machine.* In the 1930's E. V. Condon designed a special purpose computer for playing the game of *Nim.* Claud Shannon of MIT built a chess-playing machine named *Caissac* and went on to define how general purpose computers could be used to play Chess.

But why should people go to the trouble of designing game playing machines? Alan Turing, a brilliant British mathematician who worked on the early computers, was among the first to study the playing of games. Although he enjoyed playing games "just for the fun of it", he also though it might lead to important programming advances, and help in serious work in business and economics.

Turing was right; the programming of digital computers to play games has led to many important insights and to a number of new programming techniques. It has also led more or less directly to today's work on *artificial intelligence*, which is right to the forefront of computer technique.

Today one may find *Nim, Tic-Tac-Toe, Go, Magic Squares, Chess* and *Checkers, Roulette, Blackjack* and many other mathematical puzzles and games played or simulated on several different general purpose digital computers.

A digital computer may be programmed to play games in many different ways. The computer may be pitted against a human player as in a game of Tic-Tac-Toe or Chess. The computer may be asked to generate the solution to a specific game, such as a Magic Square. The computer can also be used as a bookkeeper between two players or teams as in war and business games. The most popular form of play for the computer is for it to participate in the game as an active player. In this type of play the human player indicates each of his moves to the computer on an input unit. The computer will then compute its move and output the move to the human player. This is usually done on a typewriter or line printer although punched card, cathode ray display tubes or other output units could be used. The computer always keeps score by recording both the computer's moves and the human's moves.

In games where no known algorithm exists, the computer often looks several moves ahead, examining all possible combinations of its own moves and those of the opponent, and selecting the move which is most advantageous according to some computable criterion of selecting a position. The above procedure is often used in chess and checker programs. This method somewhat simulates the action of a human player. However, there are many different, possible moves in a game of Chess and there are millions of possible situations resulting from just a few moves. Each opponent starts with 16 different pieces. With nearly all of them, he has a choice of several different moves on the board after the game has opened up, and these moves, in turn, affect the other moves open to the opponent. In an average game of Chess, each player makes 40 moves, and each move has about 30 possibilities. This results in about 10^{120} possible variations in a game. The size of this number staggers the imagination and few computers are large enough to analyze all possible sequences of moves.

There are other games that computers have not been programmed to play very well. Poker and Bridge are two of these games. They often involve *bluffing* and *counter-bluffing*, an exercise that computers are unable to participate in. A computer can, however, use a mixed strategy with pseudo random numbers in order to bluff a suitable portion of the time. The difficulty in programming a computer to play a game greatly depends upon the complexity of the game. Games with simple known algorithms, such as Magic Squares, Tic-Tac-Toe, Nim and 15 Puzzle, can be programmed without much trouble. However, programs to play games like Checkers, Chess and Go will be complex programs primarily because the game strategy is extremely complex. In Chess, where not even a chess master can outline a precise algorithm for evaluating the desirability of a certain move, it is easy to see the complexity of a chess-playing program. The end game, however, can be analyzed to a conclusion and one can write a program to evaluate this situation.

1.2 Game-Playing Programs

A few games that have been programmed for play on digital computers are identified in this section. The rules for playing these games are covered elsewhere in the book.

Tic-Tac-Toe. The idea of playing Tic-Tac-Toe on a machine was conceived as far back as the 1800's. Charles Babbage, an English mathematician, wanted to build a machine to play Chess and Tic-Tac-Toe to help finance his efforts to build his Analytical Engine.

Today one may find special purpose machines that play Tic-Tac-Toe in such a common place as Disneyland in Southern California. Tic-Tac-Toe programs have been written for many digital computers.

Go. The Japanese game of Go is an extremely popular game among computer people. The game is played with black and white stones on a board containing 361 intersection points. The object of the game is to surround vacant intersection points.

The rules of Go are simple and no mathematical theory of the game is known. It is estimated that there are around 10^{172} different board positions during the course of a game. It is easily seen that it would be impossible to calculate all the various board configurations during the course of a game. This is one of the reasons that Go is such an interesting game to play on a computer.

Several Go programs have been written for digital computers. One such program consisted of around 30,000 instructions and data words on an IBM-704 computer.

Go is probably the most difficult of the board games. No really successful Go-playing program has yet appeared. However, Thorp and Walden (1970) investigated some of the logical aspects of the game; Zobrist (1969) described a program that plays a legal game and has "reached the bottom rung of the ladder of human Go players," and Ryder (1971) described a program that uses heuristic search techniques to play a "fair beginner's" game. It may be several years before a program can be written that will be "skillful" at playing the game.

Pentominoes. A polyomino is a figure formed by joining unit squares along their edges. Pentominoes are five-square polyominoes and it is possible to construct 12 different pentominoes. A pentomino game is played by arranging the 12 pentominoes into various size rectangular boxes: 3 by 20, 4 by 15, 5 by 12, or 6 by 10. Computers have been used to generate many solutions to the pentomino game. In fact, a computer program found that there are two solutions for the 3 by 20 configuration, 1010 for the 5 by 12 configuration and 2339 for the most popular size, the 6 by 10 rectangular configuration.

Knight's Tour. The strange moves of the chess *knight* make his operations fascinating. He is permitted to move two or one rows up or down and one or two columns left or right on the chessboard. An interesting game is to move the knight to every square on the chessboard without landing in any square twice. This game is called a Knight's Tour. There are many different tours and digital computers have been used to determine many of them.

Go-Moko. Go-Moko is a two-player game played on a 19 by 19 lined Go board. Each player has 180 stones and places the stones, on alternate moves, on an intersection of the board. The object is to obtain five adjacent stones in a row either vertically, horizontally, or diagonally. The player doing this wins the game. Several computer programs have been written to play this game.

Wolf, Goat and Cabbage Puzzle. Many problems have been designed using odd combinations of people or objects which have to be moved across an imaginary river. The Wolf, Goat and Cabbage Puzzle is one of these problems.

A man wants to cross a river taking with him a wolf, a goat and a basket of cabbages in a boat that can hold only himself and one of the three at a time. If left alone, the wolf would eat the goat, and the goat (if not yet devoured by the wolf) would eat the cabbages. The puzzle involves ferrying the precious cargo across the river without a catastrophe occurring.

When solving this puzzle on a computer, one must represent the conditions with numbers instead of words. For example, assign the following numbers to the man, wolf, goat and cabbages.

MAN	1000
GOAT	111
WOLF	10
CABBAGES	1

Any time the numbers on either riverbank add up to a total that is more than 111 and less than 1011

$$\text{TOTAL} < 1011$$
$$\text{TOTAL} > 111$$

the solution proposed by the program is wrong and the computer must try another combination.

Problems of this type, even the much more complicated ones, are relatively simple for a computer to solve. The computer simply looks at all possible conditions until it finds a correct solution.

15 Puzzle. The 15 Puzzle consists of a square box containing squares with the numbers 1 to 15 and one blank square. Any one of the numbers to the immediate right, left, top, or bottom of the blank square can be moved into the blank space. The object of the puzzle is to start with a specific number arrangement and finish with a different arrangement. There is one slight catch to the puzzle—there are 10,461,394,944,000 number arrangements that are impossible to obtain. There are also the same number of possible arrangements.

A computer program of around 100 machine language statements can determine if a specified number arrangement of the 15 Puzzle is possible or impossible.

Nim. The ancient mathematical game of Nim has always been a favorite of computer people. Prior to 1945, several machines were built to play Nim. Since the invention of the electronic digital computer, many computer programs have been written to play this game.

Nim is played by two people or one person and a computer playing alternately. Before the play starts, an arbitrary number of objects is put in an arbitrary number of piles, in no specific order. Then each player in his turn removes as many objects as he wishes from any pile (but from only one pile and at least one object). The player who takes the *last* chip is the winner of the game.

The game is thought to be of Chinese origin; however, the name Nim was assigned to the game in 1901 by Charles Bouton, a professor of mathematics at Harvard University. This was the first time that a complete analysis of the game had been conducted. The algorithm for Nim removes the element of chance from the game. A player knowing the algorithm can always improve his chances of winning the game. Likewise, a computer program designed to play Nim properly can usually win against a player not knowing the algorithm. The winning secret to playing Nim is to always present your opponent with an *even* position. An *even* position is determined (1) by writing the number of objects in each pile in binary notation; (2) obtaining the sum of the digits of every column of the binary numbers; and (3) dividing the obtained sum by 2. The position is even if no remainder resulted from the division.

Prime Numbers. An integer greater than one is called a Prime Number if and only if the only positive integers that exactly divide it are itself and the number one. The prime numbers less than 25 are 2, 3, 5, 7, 11, 13, 17, 19 and 23. How does one determine if a number is prime? One way is to write down a large number of integers and simply cross off the composite numbers (numbers that are divisible by numbers other than themselves and the number 1). This simple procedure was devised by Eratosthenes 2000 years ago. This procedure is relatively easy to use when one wants to determine only a few Prime Numbers; however, it would be a rather lengthy operation to determine all the primes less than 200,000 or to determine if 209267 is a prime number. A computer can easily determine if a number is prime by using a method similar to that of Eratosthenes. A computer was used to determine a 961-digit prime number ($2^{3217}-1$), that 1,000,000,009,649 and 1,000,000,009,651 were twin primes (prime numbers with a difference of 2) and that $2^{11213}-1$ was a 3376-digit Prime Number.

War Games. War Games have been used by military organizations for many years. Computers have been used in gaming theory since about 1950. Computers are used in War Games to evaluate military strategies, weapon systems, tactics and organizational concepts. They are used to simulate activity ranging from the small military unit tactical action to a large full-scale war using many armies.

As an example of a recent computer war game, consider *Grand Strategy*, which is a game of international conflict developed by Raytheon Company for the U. S. Department of Defense. The global cold war conflict incorporates three power alliances of 39 nations with conflicting interests. The action takes place over a simulated ten-year period, divided into weekly events. In this game, all nations can win peace and prosperity. The roles of the players are as political, military and economic leaders of the nations.

Checkers. Checkers can be played between a human player and a computer. The checkerboard is divided into 64 squares, colored alternately light and dark, and each side is provided with twelve men, known as *white* and *black*. At the beginning of the game the board is so placed that each player shall have two of his men touching the edge of the board at his left. The men never leave the color upon which they are first placed, and all moves must be diagonal. The object of the game is to capture all the opponent's men and remove them from the board, or else pen them up in such a manner that they cannot move.

The game of Checkers is somewhat simpler than Chess since it involves simpler moves, 32 instead of 64 playing squares, and about eight instead of more than 30 possible moves for an average chess position. Still, it has been estimated that there are around *10 duodecillion*

$$10,000,000,000,000,000,000,000,000,000,000,000,000,000$$

possible moves in an average game.

Several checker-playing computer programs have been written; however the most important program was written by Dr. Arthur Samuel. Dr. Samuel's checker-playing program improves its behavior in terms of games won as it accumulates experience. It can learn in two ways: 1) a rote method of remembering past board positions and corresponding evaluations; and 2) a scoring method in terms of particular questions asked about the board (such as, will this move block a jump? or, will this move get a King?). Several dozen such questions are assigned arbitrary initial values and are devalued or envalued according to their participation in unsuccessful or successful moves. That is, a move is selected if, according to the sum of the scoring questions, it has the highest value. When a bad move is made (e.g., a piece is lost on the next move), all contributing scoring questions are devalued. Only 16 of the several dozen questions are used at a time and the least valuable question is periodically removed from activity and replaced by the question that has been inactive the longest.

Dr. Samuel's program is capable of playing Checkers at a championship level. The program is capable of beating all but the very best players, and once beat a Checkers master, Robert W. Nealey (see Figure 1-3).

Game move	Black (computer)	White (Nealey)	Game move	Black (computer)	White (Nealey)
1	11—15		28		27—23
2		23—19	29	15—19	
3	8—11		30		23—16
4		22—17	31	12—19	
5	4—8		32		32—27
6		17—13	33	19—24	
7	15—18		34		27—23
8		24—20	35	24—27	
9	9—14		36		22—18
10		26—23	37	27—31	
11	10—15		38		18—9
12		19—10	39	31—22	
13	6—15		40		9—5
14		28—24	41	22—26	
15	15—19		42		23—19
16		24—15	43	26—22	
17	5—9		44		19—16
18		13—6	45	22—18	
19	1—10—19—26		46		21—17
20		31—22—15	47	18—23	
21	11—18		48		17—13
22		30—26	49	2—6	
23	8—11		50		16—11
24		25—22	51	7—16	
25	18—25		52		20—11
26		29—22	53	23—19	
27	11—15			White concedes	

Figure 1-3. A victory for Dr. Arthur Samuel's checker playing program.

Dr. Samuel's checker playing program is based on the rote method of learning, as illustrated in Figure 1-4.

Chess. The first machine that could play Chess was a *fake*. It was late in the 18th century (1769) when a Hungarian inventor named Wolfgang von Kempelen introduced a game-playing machine known as the *Maelzel Chess Automation.* Von Kempelen toured Europe with this *automatic machine* showing it to large audiences. It was later found out that von Kempelen's Chess player was a clever fake, and was not an automatic

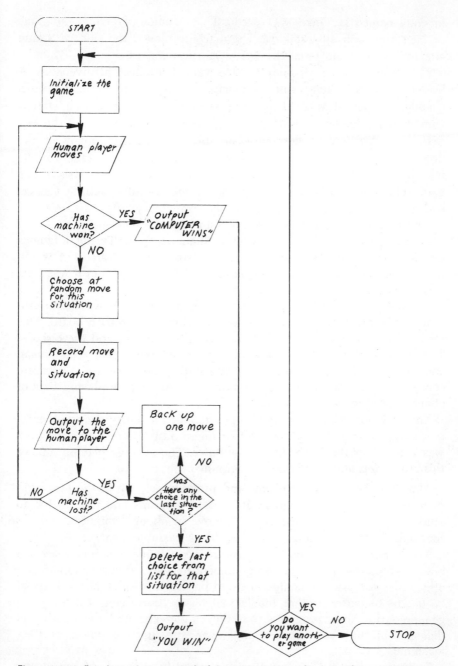

Figure I-4. A flowchart of a rote method for a computer to learn to play a winning game.

machine at all. In fact it played Chess through the direction of a chess-playing midget who was concealed inside the cabinet.

In 1914 an honest chess-playing machine was invented by a Spanish inventor named L. Torres y Quevedo. This machine played an end game of king and rook against king. The machine played the side with king and rook and would force checkmate in a few moves.

The game of Chess is one of man's valued intellectual diversions. A leader in the development of chess-playing computer programs has been Claude Shannon of MIT. Shannon, a skilled chess player himself, started game-playing history with a paper in 1949. In this paper he did not describe a particular chess program, but he discussed many of the problems involved with using game-playing machines. The framework he introduced established a solid foundation for subsequent analysis of this game. Shannon also built a chess-playing machine called *Caissac*. Caissac had a relay memory that contained the elementary rules of Chess and played simple end games where it had an advantage of one piece.

In 1950, following the lead of Claude Shannon, A. M. Turing, a famous English mathematician and logician, described a program that would play Chess. Turing's program played on an 8 by 8 board, but it was later classified as a very weak chess player.

In 1956, a group of mathematicians at Los Alamos, New Mexico programmed a MANIAC I computer to play Chess on a 6 by 6 board. This resulted in eliminating special chess moves, four pawns and the bishops. The playing ability of this program was rated equivalent to a human chess player with an experience of 20 games. The MANIAC I program consisted of around 600 instructions and had an average playing time of about 12 minutes per move.

The same Los Alamos group also programmed a chess-playing program that played a complete game of Chess on an 8 by 8 board. This program was written for the MANIAC II computer. The average playing time of this program is around 18 minutes per move.

Alex Bernstein, a chess player and programmer, wrote the first complete chess-playing program in 1957. The program was written for an IBM-704 computer and can make a move in about 8 minutes. It looks four half-moves in advance, and plays like a passable amateur.

Nearly twenty-five years have passed since Claude Shannon described how a computer might be programmed to play chess. Shannon and others thought that if a computer could be taught to play chess, it could be taught to perform other intellectual tasks. Researchers excitedly began preparing their programs, but they underestimated the depth and difficulty of the chess problem and over-estimated the power of their machines.

These early frustrations have to some extent been eased. And what was theory in the early fifties is an annual tournament now. The first nationally organized ACM* computer chess tournament took place in New

---

* Association for Computing Machinery, a professional computer organization.

York City in August 1970. Three years and three tourneys later, the electronic chess masters met in Sweden for the first international face-off.

In 1974, computer chess had its first world champion. After four years of domination of computer chess in the ACM sponsored U.S. Computer Chess Championships, CHESS 4.0 of Northwestern University fell to defeat at the teletypewriter of CHAOS from Univac (see Figure 1-5). It was the program KAISSA from the Soviet Union which swept the tournament with a perfect score of four points in the four-round tournament (see Figure 1-6).

White	Black	White	Black
1 P-Q4	P-Q4	25 N-B5	B-B4
2 P-QB4	PxP	26 P-N4	Q-K1
3 N-KB3	N-KB3	27 B-R4 (h)	P-N6
4 P-K3	P-K3	28 PxB	PxP
5 BxP	P-B4	29 BxN	P-R8=Q
6 Q-K2	P-QR3	30 RxQ	R-R3
7 O-O	P-QN4	31 NxR	Q-Q1
8 B-N3	B-N2	32 K-B2 (i)	K-B2
9 R-Q1	N/1-Q2	33 Q-K6ch	K-B1
10 N-B3	B-Q3	34 QxNch	QxQ
11 P-K4	PxP	35 RxQ	KxR
12 NxP/4	Q-N1	36 N-B5	R-QN1
13 P-N3	P-N5 (a)	37 RxP	RxPch
14 N-R4	BxP/5 (b)	38 K-N3	P-N3
15 P-B3	B-N3 (c)	39 PxP	PxP
16 NxP!(d)	PxN	40 R-R6	R-QB7
17 QxP/K6ch	B-K2	41 R-K6ch	K-B1
18 R-K1	Q-Q1	42 R-K5	R-B8
19 B-KB4!(e)	K-B1	43 R-N5	K-B2
20 R/R-Q1	R-R2	44 B-K6ch	K-B3
21 R-QB1?!(f)	N-KN1	45 P-KR4	RxN
22 R/B-Q1	P-QR4 (g)	46 RxR	KxB (j)
23 B-Q6	BxB		etc.
24 QxBch	N-K2		

Figure 1-5. Round 2 of First World Computer Chess Championship at Stockholm, Sweden, 1974. White: CHAOS; Black: CHESS 4.0.

White	Black	White	Black
1 N-KB3	P-K3	35 R-Q1?(i)	Q-N3
2 P-Q4	N-KB3	36 R-QN1	R-QB1
3 B-N5	P-Q4	37 B-K6	R-Q1
4 P-K3	B-K2	38 Q-N6 (j)	Q-N2
5 N-QB3	B-N5 (a)	39 Q-KB5??(k)	Q-QB2
6 BxN?!(b)	BxN(c)	40 R-R4	N-Q5
7 PxB	QxB	41 Q-R3	NxB
8 B-Q3	P-B4	42 QxN (l)	B-Q6
9 O-O	O-O	43 R-KN1	B-B5 (m)
10 Q-Q2	N-B3	44 Q-B5	B-K7
11 PxP (d)	Q-K2	45 R-R1	P-QR4
12 P-QB4!(c)	PxP	46 Q-N6	P-R5 (n)
13 BxP	QxP	47 R-K1	B-B5
14 Q-Q3	R-Q1	48 R-QR1	P-R6
15 Q-K4	P-QN4?(f)	49 R-QN1	Q-Q3
16 B-Q3	P-KB4	50 QxQ	RxQ
17 Q-KR4	P-K4	51 R-KR3	P-R7
18 P-K4	P-B5	52 R-QB1	R-Q5(o)
19 R(B1)-K1 (g)	B-N2??(h)	53 R(R3)-QB3	RxP
20 N-N5	P-KR3	54 R-R1	R-Q5
21 N-K6	Q-N3	55 RxB (p)	RxR
22 NxR	RxN	56 P-N3	P-B6
23 P-R4	P-N5	57 P-KR3	R-QB7
24 B-B4ch	K-R1	58 R-Q1	R-Q7
25 QR-Q1	N-Q5	59 R-QB1	P-K5
26 R-QB1	B-B3	60 P-N4	P-K6
27 P-QB3	PxP	61 K-N1	P-K7
28 RxP	BxP/R5	62 K-B2	R-Q8
29 Q-K7	N-B3	63 R-B8ch	K-R2
30 Q-KB7	Q-B4	64 KxP(B3)	P-K8=Q
31 R-Q3	N-Q5	65 R-QB2	R-Q6ch
32 B-Q5	B-N4	66 K-KB4	P-N4ch
33 R-KR3	N-K7ch	67 K-KB5	R-KB6
34 K-R1	QxP		mate

Figure 1-6. Round 4 of the Stockholm Tournament. White: OSTICH; Black: KAISSA.

The tournament site was in central Stockholm. The games were analyzed for the audience by David Levy, the tournament director, on special display boards supplied by the Swedish Chess Federation. At each teletypewriter, Cathode Ray Tube display terminal, or telephone was a flag showing the competing country. Thirteen programs were entered in the tournament and were run on computers located all over Europe.

Blackjack. Blackjack *(Twenty-One)* is the most popular card game found in casinos throughout the world. Contributing to this popularity is the fact that the game is played at a rapid rate, and money can be exchanged quickly. It is a simple game to play once all the basic rules are known. For a specific combination of the player's two up cards and the dealer's one up card, there is only one course of action for the player. In principle, it is easy for a digital computer program to use the same logic.

Digital computers have been playing Blackjack since 1954. Most of these computers were located in scientific laboratories. Programs such as Blackjack present a challenge to the computer scientist and very often the knowledge gained in figuring out how to program it has a direct carry-over to more practical problems.

The first major blackjack program was written at the Atomic Energy Commission's laboratory at Los Alamos, New Mexico. It was programmed for an IBM-701 computer in 1954.

Until 1961, it was not generally known how to overcome the house odds at Blackjack. Then Dr. Edward O. Thorp wrote the book, *Beat The Dealer.* This book, which at one time was on the best seller list of Time magazine, included a basic strategy for playing Blackjack properly. Thorp's system was based on the fact that the odds change after certain cards have been played. Playing his system in a Las Vegas casino, Thorp won $2000 in about four hours. He was soon barred from several blackjack games he patronized because his winnings were too consistent. Thorp had used a computer to work out all the odds changes as certain cards were removed from the deck and got from the computer the percentages for or against the player in each situation.

In the mid-1960's, Dr. Allen Wilson, a research scientist in San Diego, California used a computer to compute blackjack statistics. He also wrote an excellent book on Blackjack and other casino games titled *The Casino Gambler's Guide.*

In 1973, Laurence Revere, a Blackjack expert, wrote the book titled *Playing Blackjack as a Business.* The strategies in this book were devised with computers by Julian H. Braun, who also wrote the original strategy detecting program for Dr. Thorp. It required more than 9,000,000,000 computer dealt hands to devise the strategies in Revere's book.

Roulette. Roulette is a popular game among computer people. Since the

American roulette wheel has 38 possible results (the numbers 1 through 36, 0 and 00), it is a good random number generator modulo 38. Computer programs have been written for several IBM and Univac computers to simulate this fascinating casino game.

The logic used in roulette programs may be applied to many practical areas. One such program written in FORTRAN uses roulette logic to calculate the effectiveness of a new shielding material (what proportion of sub-atomic particles it would screen out, absorb, or allow to pass through). A run of the program generated a mesh of 72 white atoms and 648 black atoms, generated where a particle entered the mesh, generated subsequent moves, made the moves, noted when a particle was absorbed, was reflected, or passed through the far side, and printed the results. In seven and a half minutes the computer simulated 7,600 particles and made 91,476 moves. When working with complex problems of this type, both the computer and the roulette logic (or Monte Carlo techniques) are excellent tools.

Magic Squares. **Magic Squares were known to the ancients and were thought to possess mystic and magical powers because of their unusual nature. These magical squares have little practical value; however, they provide stimulating problems for programmer training.**

Business Games. Game playing is being applied to business management. Business executives are playing games which simulate the operation of their businesses. One such game permits several teams of players to compete for a market. The teams allocate their capital to sales effort, research and development, and production each month. They select certain sales options and research options. Periodically, these options are entered into a computer as data and the teams compete, on the basis of their decisions, for the available market for the item the companies are producing. The players receive a computer output at the end of each period and thus are able to learn about the process of allocating resources and managing a business enterprise.

Twenty Questions. At an IBM display room in New York, a computer programmed to play the game "20 Questions" dared a visitor to guess its mystery work (*knowledge*). As the computer computed and the player typed on an input unit, the "conversation" went like this:

Player: Does it grow?
Computer: To answer that question might be misleading.
Player: Can I eat it?
Computer: Only as food for thought.
Player: Do computers have it?
Computer: Strictly speaking, no.

```
MACHINE --- THE ONE-ARMED BANDIT.

EACH PLAY YOU CAN PUT ANY AMOUNT OF SILVER DOLLARS INTO
THE MACHINE, UP TO THE BALANCE YOU HAVE, BY TYPING IN
THE QUANTITY OF DOLLARS.

YOU 'PULL DOWN THE HANDLE' BY DEPRESSING THE 'RETURN' KEY.

WHEN YOUR BALANCE REACHES ZERO, THE GAME IS OVER.....
OR TYPE 'STOP' TO QUIT AT ANY TIME. GOOD LUCK...

WHAT IS YOUR LUCKY NUMBER TODAY, BETWEEN 1 AND 711? 444

GOOD.  YOUR STARTING BALANCE IS $ 32
HOW MANY DOLLARS DO YOU WANT TO PUT IN ON YOUR FIRST PLAY?  1

ORANGE   ORANGE   ORANGE   YOU WIN: $ 10     BALANCE: $ 41    AGAIN? 22

ORANGE   PLUM     LEMON    ...A LOSER        BALANCE: $ 19    AGAIN? 1

BELL     LEMON    LEMON    ...A LOSER        BALANCE: $ 18    AGAIN? 1

CHERRY   ORANGE   ORANGE   ...A LOSER        BALANCE: $ 17    AGAIN? 1

BELL     PLUM     LEMON    ...A LOSER        BALANCE: $ 16    AGAIN? 2

ORANGE   LEMON    LEMON    ...A LOSER        BALANCE: $ 14    AGAIN? 2

ORANGE   LEMON    ORANGE   ...A LOSER        BALANCE: $ 12    AGAIN? 2

ORANGE   ORANGE   LIBERTY  YOU WIN: $ 32     BALANCE: $ 42    AGAIN? 3
```

Figure 1-7. Slot machine simulated by computer.

Slot Machines. Charles Fey, a young mechanic who emigrated from Bavaria to San Francisco in 1887, invented the Slot Machine and put it into operation in the Bay area in 1895. The first one was called "The Liberty Bell," accepted and paid out nickels, and sat on the bar in a saloon. Today there are thousands of these mechanical coin grabbers operating in casinos throughout the world. There is also another type of Slot Machine called the *computer simulated slot machine.* Instead of pulling the handle as one would do on a real Slot Machine, the action is started by pointing a light pen at a start position on a display console. The computer generates a three-symbol combination composed of the following symbols: cherries, oranges, melons, bars, bells, lemons, and plums. This symbol combination, along with an indicated payoff, is displayed on the cathode ray tube of the display console.

Slot Machines are also simulated by typewriters connected to computers. Figure 1-7 illustrates the printout of Slot Machine play.

Computers are also being used in Las Vegas casinos to monitor the operation of Slot Machines. The computer system provides a printed listing of the slot's identification, the money invested in the machine by slot enthusiasts and the amount of payoff. The computer can easily keep track of the operation of several hundred Slot Machines.

Poker and Bridge. A program developed by A. I. Wasserman (1970) is capable of bidding skillfully in the game of Contract Bridge. Bridge bidding is a significant intellectual task, involving imperfect information and requiring an ability to work and communicate with a partner. Wasserman's program achieves the level of human experts in partnership bidding and is estimated to be slightly more skillful at competitive bidding than is the average duplicate Bridge player. The program is capable of bidding skillfully according to four systems: Standard American, Goren, Schenken, and Kaplan-Schweinwold.

In 1968, D. A. Waterman designed a language in which heuristics for Draw Poker could be expressed as sentences, and he attempted to construct a program that could select the appropriate sentences under the guidance of experience. Waterman's Poker-playing program is capable of "learning" to play a fair game of Draw Poker.

Football. Football is probably the most popular simulated sport game. Several simulation programs exist that use standard professional football rules without penalties. The computer takes the part of your opposing team and also the referee. One program allows you to run eight plays on offense and five on defense. The programs present the necessary rules as you play.

Hexapawn. The game of Hexapawn and a method to learn a strategy for playing the game was described in "Mathematical Games" in the March 1962 issue of *Scientific American.* The method described in the article was for a hypothetical learning machine composed of match boxes and colored beads. Several computer programs have been written to play Hexapawn.

Russian Roulette and Other Games. In a game program called RUSROU, you are given by the computer a revolver loaded with one bullet and five empty chambers. You spin the chamber and pull the trigger by inputting a "1," or, if you want to quit, you input a "2." You win if you play ten times and are still alive. Let's play this game.

THIS IS A GAME OF RUSSIAN ROULETTE

HERE IS A REVOLVER
PRESS "1" TO SPIN CHAMBER AND PULL TRIGGER. PRESS "2" TO GIVE UP.

GO?   1
-CLICK-
?   1
-CLICK-
?   1
-CLICK-

```
?  1
-CLICK-
?  1
-CLICK-
?  1
-CLICK-
?  1
-BANG- YOU'RE DEAD!
```

There are literally hundreds of games that have been programmed for computer play. Many are discussed in this book. Others may be found in *Game Playing with BASIC* by D. D. Spencer (Hayden Book Co.), *BASIC Computer Games* by D. H. Ahl (Digital Equipment Corp.), *Fun and Games with the Computer* by E. R. Sage (Entelek, Inc.), and other books listed in the Bibliography.

CHAPTER 2

Magic Square Programs

MAGIC SQUARES IS ONE of the oldest and most fascinating of all number curiosities. A magic square consists of an array of numbers arranged in the form of a square, so that the sum in every column, in every row and in both main diagonals is identical. The square is said to be of *order n* if there are *n* number of cells on one of its sides: order 3 squares have 3 cells to a side, order 4 squares have 4 cells to a side, order 5 squares have 5 cells to a side, etc. The sum of any column, row or main diagonal is equal to a value called the *magic number*. If the magic square is composed of the numbers 1 through n^2, then the magic number may be calculated from the formula

$$\text{magic number} = \frac{n(n^2+1)}{2}$$

where n is equal to the size of the square. If the magic square is composed of numbers other than 1 through n^2, say p through n^2+p-1, then the magic number can be found from the formula

$$\text{magic number} = \frac{n^3+n}{2} + n(p-1)$$

where p is the starting value of the magic square and n is the size of the square.

There are many simple methods which may be used for constructing magic squares of a specific order. There is only one magic square of order 3, though eight forms may be obtained through 90°, 180° and 270° rotations and mirror reflections. There are 880 squares of order 4 (7040 if all reflections and rotations are considered) and an estimated value of over 13,000,000 order 5 magic squares.

Magic squares with an odd number of cells (order 3, order 5, order 7, order 9, order 11, etc.) are usually constructed by methods which differ

23

from those governing the construction of squares with an even number of cells on a side (4, 6, 8, 10, 12, 14, etc.). Several methods for generating these number curiosities are now presented.

2.1 Odd Order Magic Square

The *De la Loubere* procedure is used to generate any magic square of odd order. For the sake of simplicity, a 5 by 5 magic square is generated in the following illustrations. The reader should keep in mind that this method of construction may be used for generating 3 by 3, 9 by 9, 11 by 11, etc., magic squares.

1. Place the number 1 in the center box of the first row.

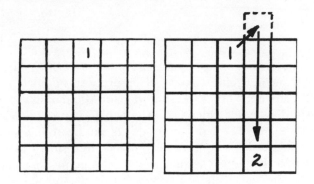

2. Move in an oblique direction, one square to the right and one square above. This movement results in leaving the top of the box. It is necessary to go to the bottom of the column in which you wanted to place the number. Place the number 2 in this location.

3. Now move diagonally to the right again and put the number 3 in the next box you enter.

4. If you continue diagonally to the right, you leave the box on the right side. When this occurs you must go to the extreme left of the row in which you wanted to place the number. After crossing over to the left side of the square, put the number 4 into the appropriate box.

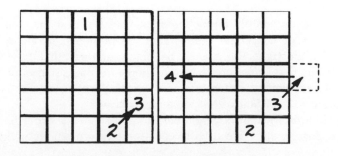

5. Now, again, go up diagonally to the right and place the number 5. This completes a group of 5 numbers.

6. Since this is a 5 by 5 magic square you must move down one box to generate the next group of 5 numbers. If this had been a 3 by 3 square or 7 by 7 square, then you would drop down when you reached a group of 3 or 7 numbers respectively.

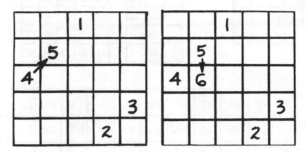

7. Move up diagonally to the right, and place a number into each box you enter. If you leave the box at the top, move to the bottom of the column where you wanted to place the number. If you land outside the box on the right side, move across to the opposite side. After each group of five numbers, go down one box to start the next group of five. When you finish the fifth group of five numbers, the number 25 will occupy the center box of the bottom row.

The sum of each of the five rows, five columns and two main diagonals is 65, and the sum of any two numbers which are diametrically equidistant from the center number is 26 or twice the center number.

17	24	1	8	15
23	5	7	14	16
4	6	13	20	22
10	12	19	21	3
11	18	25	2	9

Figure 2-1 illustrates other odd order magic squares that were constructed using the De la Loubere procedure. The reader is invited to follow the sequence of numbers to impress the De la Loubere method on his memory.

De la Loubere Magic Square Generating Program. This program, re-

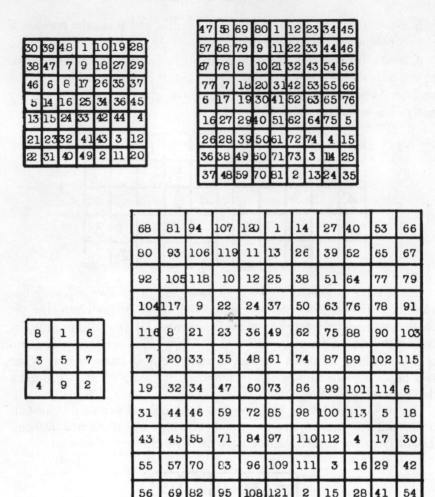

Figure 2-1. Odd order magic squares.

ferred to as the *De la Loubere Magic Square Generating Program,* can be used to generate odd order magic squares up to and including order 19 magic squares. The program could be modified to generate larger magic squares simply by increasing the arguments of the DIMENSION statement and revising the output FORMAT statement. For example, if it was desired to generate a 21 by 21 magic square, a DIMENSION statement of

DIMENSION MAGIC (21,21)

a FORMAT statement of

70 FORMAT (1HO // 21I5)

and an input card of

would cause the program to produce the desired square.

This program reads into memory an input card containing the size of the magic square that is to be generated.

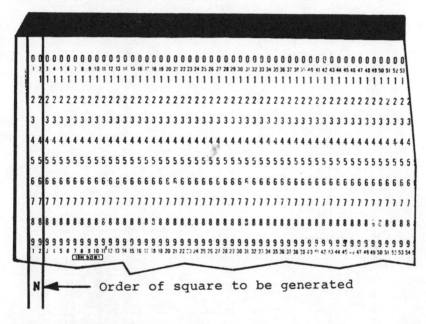

N ◄─── Order of square to be generated

The previous input data card specifies that a 13 by 13 magic square is to be generated. The program reads the card contents into memory and computes a 13 by 13 magic square.

The program reserves computer core memory for the magic square by using the DIMENSION statement. Starting values for KTR and NUM are set to 1. KTR is a program counter that is used to determine multiples of N. NUM is a variable that will vary from 1 to N^2 and it is this value that is stored in Array MAGIC.

The subscripts I and J are first set to 1 and $(N+1)/2$, respectively, which will specify the center cell of row 1 when appended to Array MAGIC. After NUM is stored in cell MAGIC(I,J), it is then incremented by 1 and compared against the largest number of the magic square, which is N^2. If the value of NUM exceeds N^2, then the program prints the magic square of order N. If the value of NUM did not exceed N^2, the program continues and a check is made to see if the KTR is a multiple of N. If it is, then KTR is reset to 1 and the row subscript is set to specify the next largest row. If KTR is not a multiple of N, then KTR is increased by 1 and the subscripts I and J are updated to specify the next cell to the *right* and *up*. If the new value of I is less than 1, which indicates that a location outside the *top* of the square was specified, the row subscript is set to address the last row of Array MAGIC. If the new

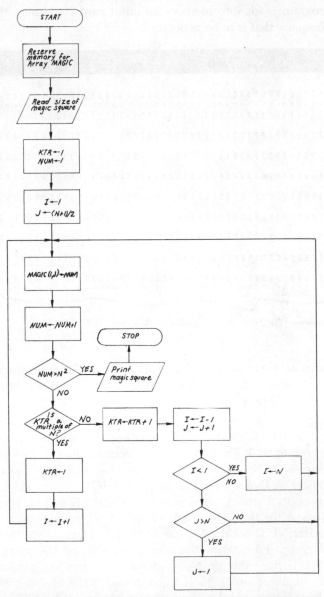

Figure 2-2. Flowchart of the De la Loubere magic square generating program.

value of J is greater than N, which indicates that a location outside the *right side* of the square was specified, the column subscript is set to address the first column of Array MAGIC. The new value of NUM is then stored in the new cell of MAGIC(I,J) and this process will continue until NUM exceeds N^2. When this occurs, the magic square has been calculated.

A flowchart of the *De la Loubere Magic Square Generating Program* is shown in Figure 2-2. The FORTRAN program is shown below.

```
C       GENERATES A MAGIC SQUARE OF SIZE N BY N

        DIMENSION MAGIC(19,19)

C       READ SIZE OF MAGIC SQUARE TO BE GENERATED

        READ 5, N

00005 FORMAT (I2)

        KTR=1

        NUM = 1

        I = 1

        J = (N+1)/2

C       PUT 1 IN CENTER CELL OF TOP ROW

00010 MAGIC(I,J) = NUM

        NUM = NUM + 1

C       JUMP TO PRINT MAGIC SQUARE IF SQUARE IS FILLED

C       CHECK TO SEE IF THE LAST NUMBER HAS BEEN PLACED

        IF (NUM - N**2)15,15,55

C       CHECK TO SEE IF KTR IS AN EVEN VALUE OF N

00015 IF(KTR - N) 30,20,20

C       RESET KTR TO 1 AND SET ROW INDEX TO NEXT ROW

00020 KTR =1

        I = I+1

        GO TO 10

C       INCREASE KTR BY 1 AND MOVE TO THE RIGHT AND UP

00030 KTR = KTR +1

        I= I-1

        J= J+1

        IF(I)40,35,40

C       SET ROW INDEX TO N (MOVE WENT OUT OF TOP OF SQUARE)

00035 I = N

        GO TO 10
```

```
00040  IF(J-N)10,10,50
C      SET COLUMN INDEX TO 1 (MOVE WENT OUT OF RIGHT SIDE OF SQUARE)
00050  J=1
       GO TO 10
C      PRINT HEADING AND MAGIC SQUARE
00055  PRINT 60,N,N
00060  FORMAT(15 H1   MAGIC SQUARE,I4,2X,3H BY ,I4)
       PRINT 70,((MAGIC(I,J),J=1,N),I=1,N)
00070  FORMAT(1H0 / 13I5)
       STOP
       END
```

Figure 2-3 illustrates a 13 by 13 magic square that was generated by the De la Loubere FORTRAN generating program.

93	108	123	138	153	168	1	16	31	46	61	76	91
107	122	137	152	167	13	15	30	45	60	75	90	92
121	136	151	166	12	14	29	44	59	74	89	104	106
135	150	165	11	26	28	43	58	73	88	103	105	120
149	164	10	25	27	42	57	72	87	102	117	119	134
163	9	24	39	41	56	71	86	101	116	118	133	148
8	23	38	40	55	70	85	100	115	130	132	147	162
22	37	52	54	69	84	99	114	129	131	146	161	7
36	51	53	68	83	98	113	128	143	145	160	6	21
50	65	67	82	97	112	127	142	144	159	5	20	35
64	66	81	96	111	126	141	156	158	4	19	34	49
78	80	95	110	125	140	155	157	3	18	33	48	63
79	94	109	124	139	154	169	2	17	32	47	62	77

Figure 2-3. A 13 by 13 magic square generated by the De la Loubere generating program.

By changing the input data card and the output FORMAT statement, various other odd order magic squares may be generated by this program. Figures 2-4, 2-5, 2-6 and 2-7 illustrate several of these.

47	58	69	80	1	12	23	34	45
57	68	79	9	11	22	33	44	46
67	78	8	10	21	32	43	54	56
77	7	18	20	31	42	53	55	66
6	17	19	30	41	52	63	65	76
16	27	29	40	51	62	64	75	5
26	28	39	50	61	72	74	4	15
36	38	49	60	71	73	3	14	25
37	48	59	70	81	2	13	24	35

Figure 2-4. A 9 by 9 magic square generated by the De la Loubere generating program.

68	81	94	107	120	1	14	27	40	53	66
80	93	106	119	11	13	26	39	52	65	67
92	105	118	10	12	25	38	51	64	77	79
104	117	9	22	24	37	50	63	76	78	91
116	8	21	23	36	49	62	75	88	90	103
7	20	33	35	48	61	74	87	89	102	115
19	32	34	47	60	73	86	99	101	114	6
31	44	46	59	72	85	98	100	113	5	18
43	45	58	71	84	97	110	112	4	17	30
55	57	70	83	96	109	111	3	16	29	42
56	69	82	95	108	121	2	15	28	41	54

Figure 2-5. A 11 by 11 magic square generated by the De la Loubere generating program.

190	191	211	231	251	271	291	311	331	351
169	189	209	210	230	250	270	290	310	330
148	168	188	208	228	229	249	269	289	309
127	147	167	187	207	227	247	248	268	288
106	126	146	166	186	206	226	246	266	267
85	105	125	145	165	185	205	225	245	265
64	84	104	124	144	164	184	204	224	244
43	63	83	103	123	143	163	183	203	223
22	42	62	82	102	122	142	162	182	202
1	21	41	61	81	101	121	141	161	181
360	19	20	40	60	80	100	120	140	160
339	359	18	38	39	59	79	99	119	139
318	338	358	17	37	57	58	78	98	118
297	317	337	357	16	36	56	76	77	97
276	296	316	336	356	15	35	55	75	95
255	275	295	315	335	355	14	34	54	74
234	254	274	294	314	334	354	13	33	53
213	233	253	273	293	313	333	353	12	32
192	212	232	252	272	292	312	332	352	11

350	9	29	49	69	89	109	129	149
329	349	8	28	48	68	88	108	128
308	328	348	7	27	47	67	87	107
287	307	327	347	6	26	46	66	86
285	286	306	326	346	5	25	45	65
264	284	304	305	325	345	4	24	44
243	263	283	303	323	324	344	3	23
222	242	262	282	302	322	342	343	2
201	221	241	261	281	301	321	341	361
180	200	220	240	260	280	300	320	340
159	179	199	219	239	259	279	299	319
138	158	178	198	218	238	258	278	298
117	137	157	177	197	217	237	257	277
96	116	136	156	176	196	216	236	256
94	114	115	135	155	175	195	215	235
73	93	113	133	134	154	174	194	214
52	72	92	112	132	152	153	173	193
31	51	71	91	111	131	151	171	172

Figure 2-6. A 19 by 19 magic square generated by the De la Loubere generating program.

231	232	254	276	298	320	342	364	386	408	430
208	230	252	253	275	297	319	341	363	385	407
185	207	229	251	273	274	296	318	340	362	384
162	184	206	228	250	272	294	295	317	339	361
139	161	183	205	227	249	271	293	315	316	338
116	138	160	182	204	226	248	270	292	314	336
93	115	137	159	181	203	225	247	269	291	313
70	92	114	136	158	180	202	224	246	268	290
47	69	91	113	135	157	179	201	223	245	267
24	46	68	90	112	134	156	178	200	222	244
1	23	45	67	89	111	133	155	177	199	221
440	21	22	44	66	88	110	132	154	176	198
417	439	20	42	43	65	87	109	131	153	175
394	416	438	19	41	63	64	86	108	130	152
371	393	415	437	18	40	62	84	85	107	129
348	370	392	414	436	17	39	61	83	105	106
325	347	369	391	413	435	16	38	60	82	104
302	324	346	368	390	412	434	15	37	59	81
279	301	323	345	367	389	411	433	14	36	58
256	278	300	322	344	366	388	410	432	13	35
233	255	277	299	321	343	365	387	409	431	12

34	57	80	103	126	128	151	174	197	220	243	266	289	312	335	337	360	383	406	429	11
56	79	102	125	127	150	173	196	219	242	265	288	311	334	357	359	382	405	428	10	33
78	101	124	147	149	172	195	218	241	264	287	310	333	356	358	381	404	427	9	32	55
100	123	146	148	171	194	217	240	263	286	309	332	355	378	380	403	426	8	31	54	77
122	145	168	170	193	216	239	262	285	308	331	354	377	379	402	425	7	30	53	76	99
144	167	169	192	215	238	261	284	307	330	353	376	399	401	424	6	29	52	75	98	121
166	189	191	214	237	260	283	306	329	352	375	398	400	423	5	28	51	74	97	120	143
188	190	213	236	259	282	305	328	351	374	397	420	422	4	27	50	73	96	119	142	165
210	212	235	258	281	304	327	350	373	396	419	421	3	26	49	72	95	118	141	164	187
211	234	257	280	303	326	349	372	395	418	441	2	25	48	71	94	117	140	163	186	209
233	256	279	302	325	348	371	394	417	440	1	24	47	70	93	116	139	162	185	208	231
255	278	301	324	347	370	393	416	439	21	23	46	69	92	115	138	161	184	207	230	232
277	300	323	346	369	392	415	438	20	22	45	68	91	114	137	160	183	206	229	252	254
299	322	345	368	391	414	437	19	42	44	67	90	113	136	159	182	205	228	251	253	276
321	344	367	390	413	436	18	41	43	66	89	112	135	158	181	204	227	250	273	275	298
343	366	389	412	435	17	40	63	65	88	111	134	157	180	203	226	249	272	274	297	320
365	388	411	434	16	39	62	64	87	110	133	156	179	202	225	248	271	294	296	319	342
387	410	433	15	38	61	84	86	109	132	155	178	201	224	247	270	293	295	318	341	364
409	432	14	37	60	83	85	108	131	154	177	200	223	246	269	292	315	317	340	363	386
431	13	36	59	82	105	107	130	153	176	199	222	245	268	291	314	316	339	362	385	408
12	35	58	81	104	106	129	152	175	198	221	244	267	290	313	336	338	361	384	407	430

Figure 2-7. A 21 by 21 magic square generated by the De la Loubere generating program.

2.2   The Agrippa Method

A simple and easy method of generating odd order magic squares is the method developed by *Agrippa*.

1. Place the number 1 in the square just below the center square.

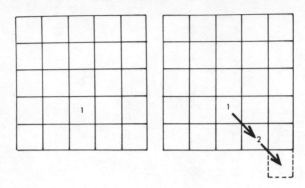

2. Place succeeding numbers on the diagonal leading to the right and down. Keep following this rule until you are outside the square.

3. When you run out of the square at the bottom, go to the top of the next column.

4. Move to the right and down and when you run out of squares on the right side of the square, go to the other end of the next row.

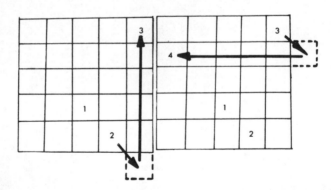

5. Move to the right and down again.

6. If the space in which you wish to place a number is already occupied by another number, put the new number two spaces below the last number.

7. Fill the remaining squares by repeating the above rules.

8. When you fill out the main diagonal going from top left to bottom right, place the next number in the extreme right of the second row; then repeat the above rules.

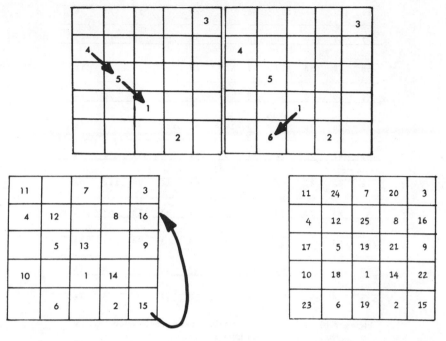

Figure 2-8, 2-9 and 2-10 illustrate magic squares that were constructed using the method of *Agrippa*.

22	47	16	41	18	35	4
5	23	48	17	42	11	29
30	6	24	49	18	36	12
13	31	7	25	43	19	37
38	14	32	1	26	44	20
21	39	8	33	2	27	45
46	15	40	9	34	3	28

Figure 2-8. A 7 by 7 magic square constructed by the Agrippa method.

37	78	29	70	21	62	13	54	5
6	38	79	30	71	22	63	14	46
47	7	39	80	31	72	23	55	15
16	48	8	40	81	32	64	24	56
57	17	49	9	41	73	33	65	25
26	58	18	50	1	42	74	34	66
67	27	59	10	51	2	43	75	35
36	68	19	60	11	52	3	44	76
77	28	69	20	61	12	53	4	45

Figure 2-9. A 9 by 9 magic square constructed by the Agrippa method.

Agrippa Magic Square Generating Program. The *Agrippa Magic Square Generating Program* can be used to generate odd order magic squares up to and including order 21 magic squares. The program reads an input data card which specifies the size square that the program is to generate.

56	117	46	107	36	97	26	87	16	77	6
7	57	118	47	108	37	98	27	88	17	67
68	8	58	119	48	109	38	99	28	78	18
19	69	9	59	120	49	110	39	89	29	79
80	20	70	10	60	121	50	100	40	90	30
31	81	21	71	11	61	111	51	101	41	91
92	32	82	22	72	1	62	112	52	102	42
43	93	33	83	12	73	2	63	113	53	103
104	44	94	23	84	13	74	3	64	114	54
55	105	34	95	24	85	14	75	4	65	115
116	45	106	35	96	25	86	15	76	5	66

Figure 2-10. A 11 by 11 magic square constructed by the Agrippa method.

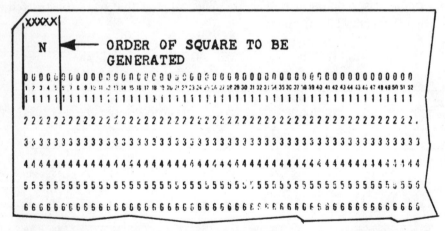

The program checks the value of N and stops the execution of the program if N is either an even number or a negative number. After determining the value of N is odd, the program then sets all locations of Array MAGIC to zero. It is in this Array that the program will generate the magic square. The subscripts I and J are used to specify a cell location of Array MAGIC. I is the row subscript and J is the column subscript. I and J are initially set to locate the center cell of row $\frac{N + 2}{2}$.

$$J = \frac{N + 1}{2} \qquad\qquad I = J + 1$$

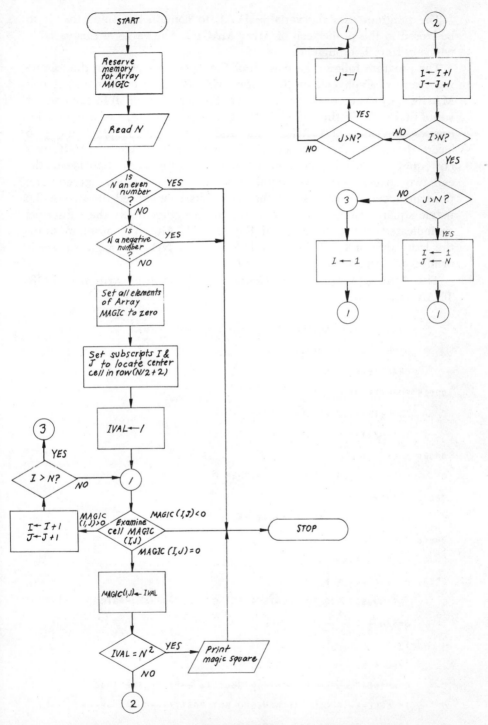

Figure 2-11. Flowchart of the Agrippa magic square generating program.

The program uses the variable IVAL to contain the value that is to be stored in the proper cell of Array MAGIC. The value of this variable will vary from 1 through N^2.

The program follows the pattern of the Agrippa method in the following manner. Each cell of the Array MAGIC is examined and if cell MAGIC (I,J) is zero, the program sets this location equal to the current value of IVAL. If the value of IVAL is less than N^2, the program increases the value of IVAL by one and sets the subscripts I and J to locate the cell to the *right* and *down*. If the cell of Array MAGIC was not equal to zero, the program sets the subscripts I and J to locate the cell two spaces below where the last value of IVAL was stored. The program logic checks to see if the subscripts ever exceed the size of the magic square being generated. If so, the program resets the subscripts as indicated in the flowchart of Figure 2-11. After the program stores the last value of IVAL in Array MAGIC, the N by N magic square is printed and program execution terminates.

The *Agrippa Magic Square Generating Program* as written in FOR-TRAN is as follows.

```
C       MAGIC SQUARE GENERATOR
        DIMENSION MAGIC(21,21)
        READ 10,N
00010 FORMAT(I5)
C       STOP IF N IS AN EVEN NUMBER
        IF((N/2)*2-N)30,20,20
00020 STOP
C       STOP IF N IS A NEGATIVE NUMBER
00030 IF(N)20,20,40
C       SET ALL ELEMENTS OF ARRAY MAGIC TO ZERO
00040 DO 50 I=1,N
        DO 50 J=1,N
00050 MAGIC(I,J) = 0
C       SET SUBSCRIPTS TO LOCATE FIRST CELL OF ARRAY MAGIC
        J=N/2 + 1
        I=J+1
        IVAL = 1
C       EXAMINE CELL OF ARRAY---IF CELL IS EMPTY PLACE IVAL
C       IN CELL---IF CELL IS NONZERO BUT POSITIVE-CONTINUE---
C       IF CELL IS NEGATIVE-STOP
```

```
00060 IF(MAGIC(I,J))20,70,150
00070 MAGIC(I,J) = IVAL
C     MAGIC SQUARE IS COMPLETE IF IVAL EQUALS MAXIMUM NUMBER (N**2)
      IF(IVAL - N**2)90,80,20
C     PRINT MAGIC SQUARE
00080 PRINT 85,((MAGIC(I,J),J=1,N),I=1,N)
00085 FORMAT(///,21I5)
      STOP
C     INCREMENT IVAL AND SET SUBSCRIPTS TO LOCATE CELL TO RIGHT AND UP
00090 IVAL = IVAL + 1
      I = I + 1
      J = J + 1
      IF(I-N)130,130,100
C     ROW SUBSCRIPT GREATER THAN N
00100 IF(J-N)120,120,110
C     ADJUST SUBSCRIPTS
00110 I=2
      J=N
      GO TO 60
C     SET ROW SUBSCRIPT
00120 I=1
      GO TO 60
00130 IF(J-N)60,60,140
00140 J=1
      GO TO 60
00150 I=I+1
      J=J-1
      IF(I-N)60,60,120
      END
```

If an input data value of 21 was used, the previous program would generate an order 21 magic square. This square is shown in Figure 2-12.

An input value of 19 could cause the program to produce the 19 by 19 square which is portrayed in Figure 2-13. Figure 2-14 illustrates a 15 by 15 magic square that was generated by this program. Statement number 85 in the FORTRAN program had to be revised in order to generate the printouts of the 19 by 19 and 15 by 15 squares.

11	252	31	272	51	292	71	312	91	332	111	352	131	372	151	392	171	412	191	432	211
232	32	273	52	293	72	313	92	333	112	353	132	373	152	393	172	413	192	433	212	12
33	253	53	294	73	314	93	334	113	354	133	374	153	394	173	414	193	434	213	13	233
254	54	274	74	315	94	335	114	355	134	375	154	395	174	415	194	435	214	14	234	34
55	275	75	295	95	336	115	356	135	376	155	396	175	416	195	436	215	15	235	35	255
276	76	296	96	316	116	357	136	377	156	397	176	417	196	437	216	16	236	36	256	56
77	297	97	317	117	337	137	378	157	398	177	418	197	438	217	17	237	37	257	57	277
298	98	318	118	338	138	358	158	399	178	419	198	439	218	18	238	38	258	58	278	78
99	319	119	339	139	359	159	379	179	420	199	440	219	19	239	39	259	59	279	79	299
320	120	340	140	360	160	380	180	400	200	441	220	20	240	40	260	60	280	80	300	100
121	341	141	361	161	381	181	401	201	421	221	21	241	41	261	61	281	81	301	101	321

342	143	364	165	386	187	408	209	430	231	11	232	33	254	55	276	77	298	99	320	121
142	363	164	385	186	407	208	429	230	10	252	32	253	54	275	76	297	98	319	120	341
362	163	384	185	406	207	428	229	9	251	31	273	53	274	75	296	97	318	119	340	141
162	383	184	405	206	427	228	8	250	30	272	52	294	74	295	96	317	118	339	140	361
382	183	404	205	426	227	7	249	29	271	51	293	73	315	95	316	117	338	139	360	161
182	403	204	425	226	6	248	28	270	50	292	72	314	94	336	116	337	138	359	160	381
402	203	424	225	5	247	27	269	49	291	71	313	93	335	115	357	137	358	159	380	181
202	423	224	4	246	26	268	48	290	70	312	92	334	114	356	136	378	158	379	180	401
422	223	3	245	25	267	47	289	69	311	91	333	113	355	135	377	157	399	179	400	201
222	2	244	24	266	46	288	68	310	90	332	112	354	134	376	156	398	178	420	200	421
1	243	23	265	45	287	67	309	89	331	111	353	133	375	155	397	177	419	199	441	221
242	22	264	44	286	66	308	88	330	110	352	132	374	154	396	176	418	198	440	220	21
42	263	43	285	65	307	87	329	109	351	131	373	153	395	175	417	197	439	219	20	241
262	63	284	64	306	86	328	108	350	130	372	152	394	174	416	196	438	218	19	240	41
62	283	84	305	85	327	107	349	129	371	151	393	173	415	195	437	217	18	239	40	261
282	83	304	105	326	106	348	128	370	150	392	172	414	194	436	216	17	238	39	260	61
82	303	104	325	126	347	127	369	149	391	171	413	193	435	215	16	237	38	259	60	281
302	103	324	125	346	147	368	148	390	170	412	192	434	214	15	236	37	258	59	280	81
102	323	124	345	146	367	168	389	169	411	191	433	213	14	235	36	257	58	279	80	301
322	123	344	145	366	167	388	189	410	190	432	212	13	234	35	256	57	278	79	300	101
122	343	144	365	166	387	188	409	210	431	211	12	233	34	255	56	277	78	299	100	321

Figure 2-12. A 21 by 21 magic square generated by the Agrippa generating program.

172	353	154	335	136	317	118	299	100	281	82	263	64	245	46	227	28	209	10
11	173	354	155	336	137	318	119	300	101	282	83	264	65	246	47	228	29	191
192	12	174	355	156	337	138	319	120	301	102	283	84	265	66	247	48	210	30
31	193	13	175	356	157	338	139	320	121	302	103	284	85	266	67	229	49	211
212	32	194	14	176	357	158	339	140	321	122	303	104	285	86	248	68	230	50
51	213	33	195	15	177	358	159	340	141	322	123	304	105	267	87	249	69	231
232	52	214	34	196	16	178	359	160	341	142	323	124	286	106	268	88	250	70
71	233	53	215	35	197	17	179	360	161	342	143	305	125	287	107	269	89	251
252	72	234	54	216	36	198	18	180	361	162	324	144	306	126	288	108	270	90
91	253	73	235	55	217	37	199	19	181	343	163	325	145	307	127	289	109	271

110	290	128	308	146	326	164	344	182	1	200	38	218	56	236	74	254	92	272
291	129	309	147	327	165	345	183	2	201	20	219	57	237	75	255	93	273	111
130	310	148	328	166	346	184	3	202	21	220	39	238	76	256	94	274	112	292
311	149	329	167	347	185	4	203	22	221	40	239	58	257	95	275	113	293	131
150	330	168	348	186	5	204	23	222	41	240	59	258	77	276	114	294	132	312
331	169	349	187	6	205	24	223	42	241	60	259	78	277	96	295	133	313	151
170	350	188	7	206	25	224	43	242	61	260	79	278	97	296	115	314	152	332
351	189	8	207	26	225	44	243	62	261	80	279	98	297	116	315	134	333	171
190	9	208	27	226	45	244	63	262	81	280	99	298	117	316	135	334	153	352

Figure 2-13. A 19 by 19 magic square generated by the Agrippa generating program.

106	219	92	205	78	191	64	177	50	163	36	149	22	135	8
9	107	220	93	206	79	192	65	178	51	164	37	150	23	121
122	10	108	221	94	207	80	193	66	179	52	165	38	136	24
25	123	11	109	222	95	208	81	194	67	180	53	151	39	137
138	26	124	12	110	223	96	209	82	195	68	166	54	152	40
41	139	27	125	13	111	224	97	210	83	181	69	167	55	153
154	42	140	28	126	14	112	225	98	196	84	182	70	168	56
57	155	43	141	29	127	15	113	211	99	197	85	183	71	169
170	58	156	44	142	30	128	1	114	212	100	198	86	184	72
73	171	59	157	45	143	31	129	2	115	213	101	199	87	185
186	74	172	60	158	46	144	17	130	3	116	214	102	200	88
89	187	75	173	61	159	47	145	18	131	4	117	215	103	201
202	90	188	76	174	62	160	33	146	19	132	5	118	216	104
105	203	91	189	77	175	63	161	34	147	20	133	6	119	217
218	91	204	77	190	63	176	49	162	35	148	21	134	7	120

Figure 2-14. A 15 by 15 magic square generated by the Agrippa generating program.

2.3   General Order 4 Square

An order 4 magic square may also be constructed by using *Bergholt's* general form as seen in Figure 2-15. The alphabetical values in this square (A, B, C, D, W, X, Y and Z) may be replaced by any positive or negative numerical value.

For example, Figure 2-16 illustrates a magic square that was constructed by this method. The numerical values used in the general square were A = 14, B = 6, C = 4, D = 10, W = 2, X = 3, Y = 14 and Z = 8.

A-W	C+W+Y	B+X-Y	D-X
D+W-Z	B	C	A-W+Z
C-X+Z	A	D	B+X-Z
B+X	D-W-Y	A-X+Y	C+W

Figure 2-15. Bergholt's general form for a 4 by 4 magic square.

ROW 1
$$A - W = 12$$
$$C + W + Y = 20$$
$$B + X - Y = -5$$
$$D - X = 7$$

ROW 3
$$C - X + Z = 9$$
$$A = 14$$
$$D = 10$$
$$B + Y - Z = 1$$

12	20	-5	7
4	6	4	20
9	14	10	1
9	-6	25	6

Figure 2-16. A 4 by 4 magic square constructed from Bergholt's general square.

The Bergholt general square will also produce *symmetrical* and *pandiagonal* squares: The square will be *symmetrical* if

$$A+C = B+D \text{ and } W+Y = X-Y = Z$$

A *symmetrical* square is shown in Figure 2-17. The values of A=3, B=6, C=5, D=2, W=4, X=16, Y=6 and Z=10 were used to determine the numerical values of the square.

The *pandiagonal* square of Figure 2-18 was generated by using the Bergholt general form and the following values for A through Z: A=12, B=4, C=6, D=4, W=3, X=3, Y=3 and Z=6.

The relationships that must exist between the variables A through Z in order for the general form to produce a pandiagonal square are:

$$\frac{A-B-C+D}{2} = Z-Y = W = X$$

Bergholt Order 4 Generating Program. The flowchart in Figure 2-20 describes a program that may be used to generate either a 4 by 4 Magic square, a symmetrical square or a pandiagonal square. An input card similar to the one shown in Figure 2-19 is used to input the data.

-1	15	16	-14
-4	6	5	9
-1	3	2	12
22	-8	-7	9

9	12	4	1
1	4	6	15
9	12	4	1
7	4	12	9

Figure 2-17. A 4 by 4 symmetrical square.   Figure 2-18. A 4 by 4 pandiagonal square.

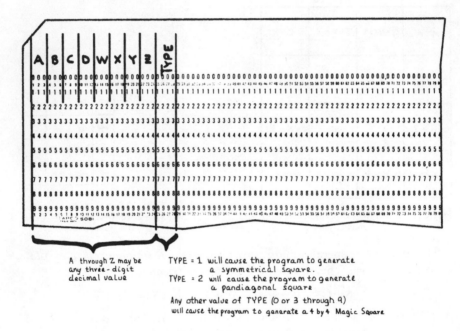

A through Z may be any three-digit decimal value

TYPE = 1 will cause the program to generate a symmetrical square.
TYPE = 2 will cause the program to generate a pandiagonal square

Any other value of TYPE (0 or 3 through 9) will cause the program to generate a 4 by 4 Magic Square

Figure 2-19. Input data card for the order 4 generator program.

The value of TYPE will determine the type of square that will be generated by the program: A symmetrical square will be generated if TYPE = 1, a pandiagonal square will be generated if TYPE = 2, and 4 by 4 magic squares will be generated for all other values of TYPE. The program checks the relationships for symmetrical and pandiagonal squares and will print an error message INCORRECT INPUT MESSAGE if the values of A, B, C, D, W, X, Y or Z do not allow the relationships to exist. If the square can be generated (the relationships are satisfied for TYPE 1 and 2), then an appropriate heading will be printed, the square values will be computed and the 4 by 4 square will be printed. An output from this program might look like that shown in Figure 2-21.

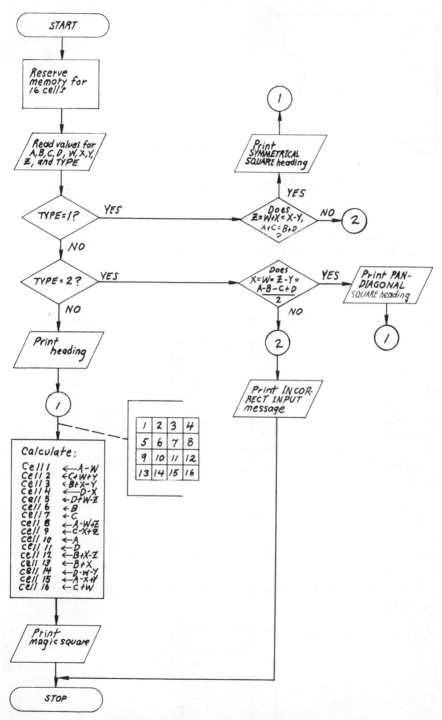

Figure 2-20. Flowchart for the order 4 generator program.

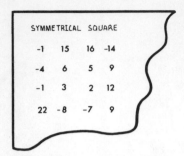

SYMMETRICAL SQUARE

-1	15	16	-14
-4	6	5	9
-1	3	2	12
22	-8	-7	9

Figure 2-21. Output square of the order 4 generator program.

2.4   Doubly-Even Order Magic Squares

Magic squares of Order 4m (squares having a multiple of 4 cells: 4, 8, 12, 16, 20, etc.) may be generated by a modified version of the Devedec Method and the Bergholt Method.

2.4.1   *Order 4 Magic Square.* The following example illustrates the generation of a 4 by 4 square.

1. In a blank 4 by 4 square, fill in the main diagonal squares with X's.

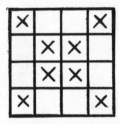

2. Start with the upper left square and move toward the right obeying the following rules:

a. If the cell is occupied by an X—skip the cell.

b. If the cell is not occupied by an X—insert a number.

Start with the number 1 and increment your count by 1 each time a move is made. When the end of a row is reached, repeat the same process in the next row.

3. The first eight numbers would be placed in the square in the following order:

	2	3	
5			8
9			12
	14	15	

Cell	X's	Contents of Square
1	X	not changed
2		2 → cell 2
3		3 → cell 3
4	X	not changed
5		5 → cell 5
6	X	not changed
7	X	not changed
8		8 → cell 8
9		9 → cell 9
10	X	not changed
11	X	not changed
12		12 → cell 12
13	X	not changed
14		14 → cell 14
15		15 → cell 15
16	X	not changed

4. Now fill in the cells containing an X. Start in the same cell as in step 2 (upper left square) and obey the following rules:

a. If the cell is occupied by an X—insert a number.

b. If the cell is occupied by a number—skip the cell.

Start with the number 16 and decrease the count by 1 each time a move is made. When the end of a row is reached repeat the same process in the next row.

5. The last eight numbers would be placed in the square in the following order:

16			13
	11	10	
	7	6	
4			1

Cell	X's	Contents of Square
1	X	16 → cell 1
2		not changed
3		not changed
4	X	13 → cell 4
5		not changed
6	X	11 → cell 6
7	X	10 → cell 7
8		not changed
9		not changed
10	X	7 → cell 10
11	X	6 → cell 11
12		not changed
13	X	4 → cell 13
14		not changed
15		not changed
16	X	1 → cell 16

6. The completed magic square would appear as follows:

16	2	3	13
5	11	10	8
9	7	6	12
4	14	15	1

ORDER 4 MAGIC SQUARE GENERATING PROGRAM. This program will generate a 4 by 4 Magic Square by a modified version of the Devedec Method.

The program reserves memory for a 4 by 4 array (Array M) and then stores 999 in each cell that lies on a diagonal and zeros in all other locations of the array. The program keeps a counter (KTR) that varies from 1 to 16 on the first pass through the array and from 16 to 1 on the second pass. The counter value is adjusted as the program progresses.

If a cell of Array M contains a zero on the first pass through the array, the counter value is placed in this cell. The value of the cell is not changed if it appears on a diagonal. On the second pass through the array, the counter value is stored in the cell if the cell contains 999. All other locations of the array remain unchanged. The calculated 4 by 4 Magic Square is printed before the program stops.

Figure 2-22 illustrates a flowchart of the program. A FORTRAN program follows the flowchart.

2.4.2 *Generating Order 8 and Order 20 Magic Squares.*

ORDER 8 MAGIC SQUARE GENERATING PROGRAM. An 8 by 8 square is constructed in a similar manner as the 4 by 4 square. The 8 by 8 square consists of four 4 by 4 squares. Figure 2-23 illustrates how the 8 by 8 diagonals are marked by placing X's on all diagonals of the 4 by 4 squares. The same procedure applies to the generation of order 8 squares as it did in the generation of 4 by 4 squares. The only difference is that the number range includes the numbers 1 through 64.

The program generates an Order 8 Magic Square by first reading into memory eight data cards that mark the diagonals. The first PRINT statement prints the marked square (Array MAGIC8). The diagonals are marked by the cells containing the value 99, as seen in Figure 2-24.

The program then places the 64 numbers in the appropriate cells of Array MAGIC8. After the last value is stored, the program prints the magic square before stopping. Figure 2-25 illustrates the 8 by 8 Magic Square that was generated by this program.

Figure 2-26 contains a printer listing of the FORTRAN program.

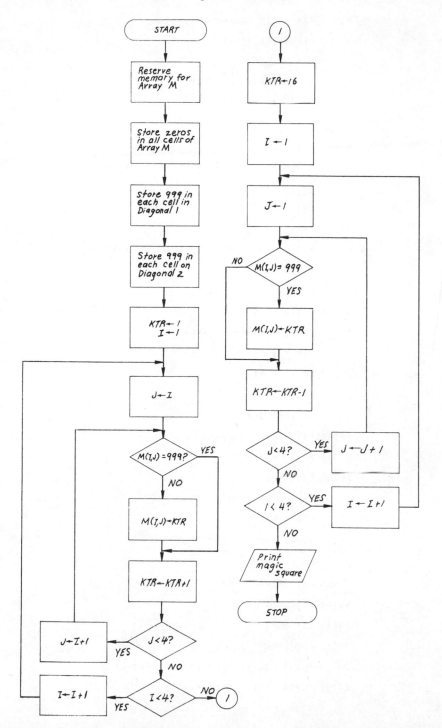

Figure 2-22. Flowchart of the order 4 modified Devedec magic square generating program.

```
C      4X4 MAGIC SQUARE GENERATOR
       DIMENSION M(4,4)
       N = 4
C      STORE ZEROS IN ARRAY M
       DO 10 I=1,N
       DO 10 J=1,N
00010  M(I,J) = 0
C      STORE 999 IN EACH CELL OF DIAGONAL 1
       DO 20 I=1,N
       J=I
00020  M(I,J) = 999
C      STORE 999 IN EACH CELL OF DIAGONAL 2
       DO 30 I=1,N
       J=N-I+1
00030  M(I,J) = 999
C      FIRST PASS THRU  ARRAY
       KTR = 1
       DO 60 I =1,N
       DO 60 J=1,N
       IF (M(I,J))50,40,50
00040  M(I,J) = KTR
00050  KTR = KTR + 1
00060  CONTINUE
C      SECOND PASS THRU ARRAY
       KTR = N*N
       DO 90 I=1,N
       DO 90 J=1,N
       IF (M(I,J) - 999) 80,70,80
00070  M(I,J) = KTR
00080  KTR = KTR-1
00090  CONTINUE
       PRINT 201,((M(I,J),J=1,N),I=1,N)
00201  FORMAT(1H1,/,(4I5,//))
       STOP
       END
```

A blank 8 X 8 square with X's on the diagonals of each 4 X 4 square.

The blank cells are filled in with numbers. Start in the upper left cell with 1 and move to the right increasing the count with each move.

The cells containing X's are filled in with numbers. Start in the upper left cell with 64 and move toward the right decreasing the count with each move.

Generated 8 X 8 Magic Square

Figure 2-23. Generating an 8 by 8 magic square.

99	0	0	99	99	0	0	99
0	99	99	0	0	99	99	0
0	99	99	0	0	99	99	0
99	0	0	99	99	0	0	99
99	0	0	99	99	0	0	99
0	99	99	0	0	99	99	0
0	99	99	0	0	99	99	0
99	0	0	99	99	0	0	99

64	2	3	61	60	6	7	57
9	55	54	12	13	51	50	16
17	47	46	20	21	43	42	24
40	26	27	37	36	30	31	33
32	34	35	29	28	38	39	25
41	23	22	44	45	19	18	48
49	15	14	52	53	11	10	56
8	58	59	5	4	62	63	1

Figure 2-24. Square with marked diagonals for the order 8 modified Devedec magic square generating program.

Figure 2-25. An 8 by 8 magic square generated by the order 8 modified Devedec magic square generating program.

```
C       8X8 MAGIC SQUARE GENERATOR

        DIMENSION MAGIC8(8,8)

C       READ DIAGONAL MARKS

        READ 10,((MAGIC8(I,J),I=1,8),J=1,8)

00010 FORMAT (32I2 / 32I2)

        PRINT 15,((MAGIC8(I,J),J=1,8),I=1,8)

00015 FORMAT(1H1,/,(8I5,//))

C       FIRST PASS THROUGH MAGIC8 SQUARE

        K=1

        DO 40 I=1,8

        DO 40 J=1,8

        IF(MAGIC8(I,J))30,20,30

00020 MAGIC8(I,J) = K

00030 K = K + 1

00040 CONTINUE

C       SECOND PASS THROUGH MAGIC8 SQUARE

        K=64

        DO 70 I=1,8

        DO 70 J=1,8

        IF(MAGIC8(I,J) - 99)60,50,60

00050 MAGIC8(I,J) = K

00060 K = K - 1

00070 CONTINUE

C       PRINT 8X8 MAGIC SQUARE

        PRINT 80,((MAGIC8(I,J),J=1,8),I=1,8)

00080 FORMAT (1H0,/,8I5)

        STOP

        END
```

Figure 2-26. A FORTRAN program for generating an order 8 magic square.

ORDER 20 MAGIC SQUARE GENERATING PROGRAM. Figure 2-27 illustrates a
FORTRAN program that will generate a 20 by 20 Magic Square. It uses
a modified version of the Devedec method for generating logic.

The 20 by 20 Magic Square that was generated by this program is
shown in Figure 2-28.

```
C       20 BY 20 MAGIC SQUARE GENERATOR

        DIMENSION KGAME(20,20)

C       READ 555 INTO EACH CELL THAT LIES ON A DIAGONAL

        READ 100,((KGAME(I,J),I=1,20),J=1,20)

00100 FORMAT (20I3)

        PRINT 150,((KGAME(I,J),J=1,20),I=1,20)

00150 FORMAT(1H1,/,(20I5,//))

C       FIRST PASS THRU ARRAY

        NUM = 1

        DO 300 I=1,20

        DO 300 J=1,20

        IF (KGAME(I,J))300,200,300

00200 KGAME(I,J) = NUM

00300 NUM = NUM + 1

C       SECOND PASS THRU ARRAY

        NUM = 400

        DO 500 I=1,20

        DO 500 J=1,20

        IF(KGAME(I,J) - 555)500,400,500

00400 KGAME(I,J) = NUM

00500 NUM = NUM - 1

C       PRINT 20 BY 20 MAGIC SQUARE

        PRINT 600,((KGAME(I,J),J=1,20),I=1,20)

00600 FORMAT (1H1,/,(20I5,//))

        STOP

        END
```

Figure 2-27. A FORTRAN program for generating
an order 20 magic square.

400	2	3	397	396	6	7	393	392	10	11	389	398	14	15	385	384	18	19	381
21	379	378	24	25	375	374	28	29	371	370	32	33	367	366	36	37	363	362	40
41	359	358	44	45	355	354	48	49	351	350	52	53	347	346	56	57	343	342	60
340	62	63	337	336	66	67	333	332	70	71	329	328	74	75	325	324	78	79	321
320	82	83	317	316	86	87	313	312	90	91	309	308	94	95	305	304	98	99	301
101	299	298	104	105	295	294	108	109	291	290	112	113	287	285	116	117	283	282	120
121	279	278	124	125	275	274	128	129	271	270	132	133	267	266	136	137	263	262	140
268	142	143	257	256	146	147	253	252	150	151	249	248	154	155	245	244	158	159	241
240	162	163	237	236	166	167	233	232	170	171	229	228	174	175	225	224	178	179	221
181	219	218	184	185	215	214	188	189	211	210	192	193	207	205	195	197	203	202	200
201	199	198	204	205	195	194	208	209	191	190	212	213	187	186	216	217	183	182	220

161	141	280	300	81	61	360	380	1
239	259	122	102	319	339	42	22	399
238	258	123	103	318	338	43	23	398
164	144	277	297	84	64	357	377	4
165	145	276	296	85	65	356	376	5
235	255	125	106	315	335	46	26	395
234	254	127	107	314	334	47	27	394
168	148	273	293	88	68	353	373	8
169	149	272	292	89	69	352	372	9
231	251	130	110	311	331	50	30	391
230	250	131	111	310	330	51	31	390
172	152	269	289	92	72	349	369	12
173	153	268	288	93	73	348	368	13
227	247	134	114	307	327	54	34	387
226	246	135	115	306	326	55	35	386
176	156	265	285	96	76	345	365	16
177	157	264	284	97	77	344	364	17
223	243	138	118	303	323	58	38	383
222	242	139	119	302	322	59	39	382
180	160	261	281	100	80	341	361	20

Figure 2-28. A 20 by 20 magic square generated by the order 20 modified Devedec magic square generating program.

2.5   Singly-Even Order Magic Squares

Magic squares of Order $2(2m+1)$ represent the 6 by 6, 10 by 10, 14 by 14, 18 by 18, 22 by 22, etc., squares. Magic squares of singly-even orders are generally the most difficult of all to construct. One method of constructing an Order 6 Magic Square is as follows:

1. Divide a 36 cell square into four 9 celled squares and label them a, b, c and d.

2. Starting with square a and the number 1 generate a 3 by 3 square.

a	b		
c	d		

8	1	6	
3	5	7	b
4	9	2	
	c		d

3. Starting with the number 10, generate a 3 by 3 magic square in square d.

4. Starting with the number 19, generate a 3 by 3 square in square b and starting with the number 28, fill the remaining 9 cells of the 6 by 6 square by generating another 3 by 3 square in square c.

8	1	6			
3	5	7	b		
4	9	2			
			17	10	15
	c		12	14	16
			13	18	11

8	1	6	26	19	24
3	5	7	21	23	25
4	9	2	22	27	20
35	28	33	17	10	15
30	32	34	12	14	16
31	36	29	13	18	11

5. Transpose the following numbers:

5 and 32

8 and 35

4 and 31

The completed square is shown on the top of the next page.

An interesting relationship of the 6 by 6 square is that the sums of the following 3 by 3 square diagonals equal the magic number of the 6 by 6 square (111).

35	1	6	26	19	24
3	32	7	21	23	25
31	9	2	22	27	20
8	28	33	17	10	15
30	5	34	12	14	16
4	36	29	13	18	11

$$\left.\begin{array}{l} 35 + 32 + 2\ + 33 + 5\ + 4 \\ 24 + 23 + 22 + 17 + 14 + 11 \\ 26 + 23 + 20 + 15 + 14 + 13 \\ 6\ + 32 + 31 + 8\ + 5\ + 29 \end{array}\right\} = 111$$

6 by 6 Magic Square Generating Program. This program will generate a magic square of Order 6. Array MAGIC is allocated COMMON storage in computer memory so both the main program and the subroutine QTRSQ may work with the array. The main program reads in a data card that specifies the size of the magic square to be generated (6). The program first prints the heading

MAGIC SQUARE 6 BY 6

where 6 is the size of the square to be generated.

A 6 by 6 Magic Square can be calculated by generating 4 smaller magic squares. This program uses the subroutine QTRSQ to generate the smaller magic squares. As seen by the flowchart of the main program, Figure 2-29, the subroutine is referenced four times—once for each quarter of the 6 by 6 square. Before transferring control to the subroutine, the main program establishes the starting value of the smaller magic square. The starting values for the four quarter squares are calculated by the main program and are stored in Array ISTART. These starting values are 1, 10, 19 and 28 and are calculated from the following equations.

$$1, (N/2)^2 + 1, 2(N/2)^2 + 1, 3(N/2)^2 + 1$$

where N is the order of the square to be calculated.

The main program establishes the order of processing the quarter squares by varying the row indicator II and the column indicator JJ.

Table 2-1 illustrates the values of II and JJ for each quarter square of the 6 by 6 square. The main program will generate the quarter squares in the order specified by the circled values of Table 2-1.

Information is passed from the main program to the subroutine by the parameters listed in Table 2-2.

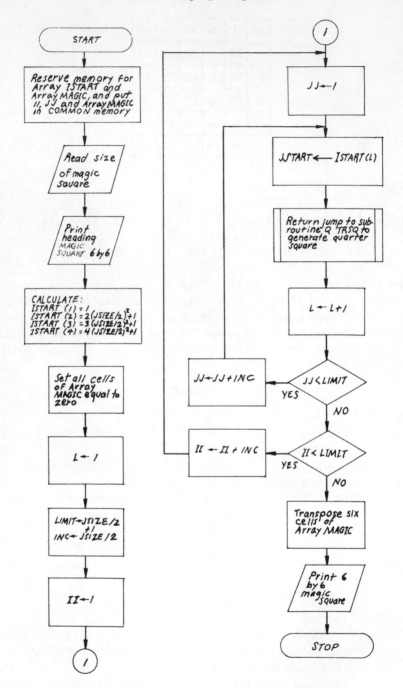

Figure 2–29. Flowchart of the 6 by 6 magic generating program.

Table 2-1. Quarter Squares of a 6 by 6 Magic Square.

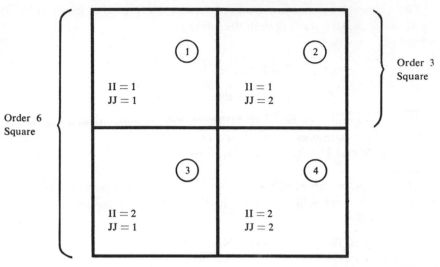

Table 2-2. Subroutine Parameters Used in the Main Program.

Main Program Parameters	II	JJ	JSTART	JSIZE
Subroutine Dummy Parameters	I	J	NUMBER	JSIZE
Program Use	Row Subscript for Quarter Square	Column Subscript for Quarter Square	Starting Value of the Quarter Square	Size of 4K+2 Magic Square

After the subroutine QTRSQ has been called on to generate the last quarter square of the magic square, the program transposes six cell locations with respect to the horizontal axis. The locations that are transposed are as follows:

MAGIC $(1,1)$ and MAGIC $(4,1)$
MAGIC $(2,2)$ and MAGIC $(5,2)$
MAGIC $(3,1)$ and MAGIC $(6,1)$

The above locations were computed from the following equations:

$$\text{MAGIC } (I,J) = \text{MAGIC } (I+B,J)$$
$$\text{MAGIC } (I+B,J) = \text{MAGIC } (I,J)$$

where the cell locations must lie on a main diagonal line, B is equal to JSIZE/2, I and J must lie in the range $1 \leq I$ or $J \leq ((JSIZE/2)/2) +1$.

After the locations are transposed, the magic square is printed.

The subroutine *Calculate Quarter Square-QTRSQ* will generate a B by B Magic Square starting with the value *s*.

where

$$B \text{ is the value } \frac{2(2k+1)}{2}$$

s is the value of JSIZE

The subroutine specifies that Array MAGIC is located in COMMON memory. The variables I and J are set equal to the row and column indicators II and JJ and are later used by the subroutine for comparison purposes.

The maximum values of the row and column subscripts and the maximum number that will be stored in the quarter square are calculated from the following equations:

MAXIMUM ROW SUBSCRIPT $N = I + 2$

MAXIMUM COLUMN
SUBSCRIPT $M = J + 2$

MAXIMUM NUMBER OF
QUARTER SQUARE $MAX = NUMBER + (JSIZE/2)^2 - 1$

A flowchart of the subroutine is shown in Figure 2-30.

The column subscript, J, is computed from the equation

$$J = J + ((JSIZE/2)/2)$$

which specifies the center column of the appropriate quarter square. Using the calculated value of the subscript J and the input value of I the first number, NUMBER, is stored in Array MAGIC.

The subroutine varies the subscripts I and J according to the flowchart in Figure 2-30, and when the last value of the quarter square (MAX) has been stored in Array MAGIC, the subroutine will return control back to the main program.

A FORTRAN program and subroutine that will calculate a 6 by 6 magic square follows:

MAIN FORTRAN PROGRAM

```
      DIMENSION MAGIC(6,6),ISTART(4)
      COMMON MAGIC,II,JJ
C     6 X 6 MAGIC SQUARE GENERATOR
C     READ SIZE OF MAGIC SQUARE (6)
      READ 10,JSIZE
```

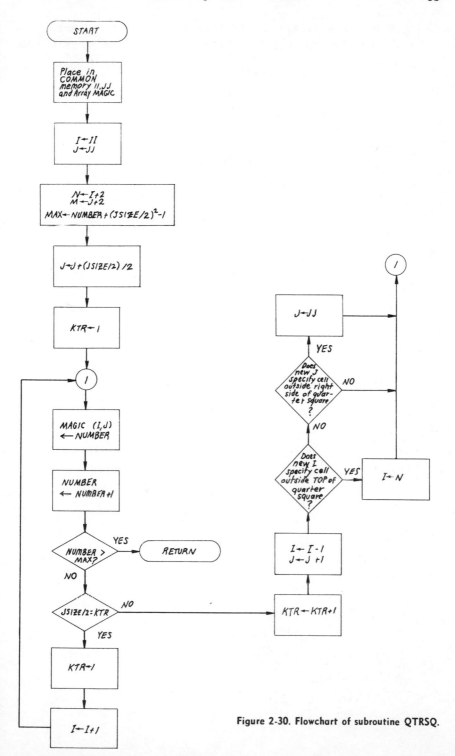

Figure 2-30. Flowchart of subroutine QTRSQ.

```
00010 FORMAT(I5)
C      PRINT HEADING
       PRINT 20,JSIZE,JSIZE
00020 FORMAT(13H1MAGIC SQUARE ,I4,2X,2HBY,I4 )
C      LOAD STARTING NUMBER TABLE
       ISTART(1) = 1
       ISTART(2) = 2 *(JSIZE/2)**2 + 1
       ISTART(3) = 3 *(JSIZE/2)**2 + 1
       ISTART(4) =(JSIZE/2)** 2 + 1
C      SET ELEMENTS OF ARRAY MAGIC EQUAL TO ZERO
       DO 25 I=1,JSIZE
       DO 25 J = 1,JSIZE
00025 MAGIC(I,J) = 0
       L = 1
       LIMIT = JSIZE / 2 + 1
       INC = JSIZE / 2
       DO 30 II = 1,LIMIT,INC
       DO 30 JJ = 1,LIMIT,INC
       JSTART = ISTART(L)
C      JUMP TO SUBROUTINE QTRSQ TO
C      GENERATE A QUARTER SQUARE
       CALL QTRSQ(JSTART,JSIZE)
00030 L = L + 1
       LL = (JSIZE / 2) / 2 + 1
       DO 40 I = 1,LL
       KK = I + JSIZE / 2
       ISTORE = MAGIC(I,I)
       MAGIC(I,I) = MAGIC(KK,I)
00040 MAGIC(KK,I) = ISTORE
       JJ = LL - 1
       DO 50 J = 1,JJ
       I = JSIZE / 2 + 1 - J
```

```
      KK = I + JSIZE / 2
      ISTORE = MAGIC(I,J)
      MAGIC(I,J) = MAGIC(KK,J)
00050 MAGIC(KK,J) = ISTORE
      PRINT 60,((MAGIC(I,J),J=1,JSIZE),I=1,JSIZE)
00060 FORMAT(1H0,6I5 / )
      STOP
      END
```

SUBROUTINE QTRSQ

```
      SUBROUTINE QTRSQ(NUMBER,JSIZE)
      DIMENSION MAGIC(6,6)
      COMMON MAGIC,II,JJ
      I = II
      J = JJ
C     CALCULATE MAXIMUM ROW,COLUMN AND NUMBER
      N = I + JSIZE / 2 - 1
      M = J + JSIZE / 2 - 1
      MAX = NUMBER + (JSIZE / 2)**2 - 1
C     SET COLUMN INDICATOR TO CENTER CELL
C     OF FIRST ROW OF QUARTER SQUARE
      J = J + ((JSIZE / 2) / 2)
      KTR = 1
00010 MAGIC(I,J) = NUMBER
      NUMBER = NUMBER + 1
C     DOES NUMBER EXCEED LARGEST NUMBER OF QUARTER SQUARE
      IF(NUMBER - MAX) 30,30,20
00020 RETURN
00030 IF(KTR - JSIZE / 2)50,40,90
C     KTR EQUALS JSIZE/2--MOVE 1 ROW DOWN
00040 KTR = 1
      I = I + 1
```

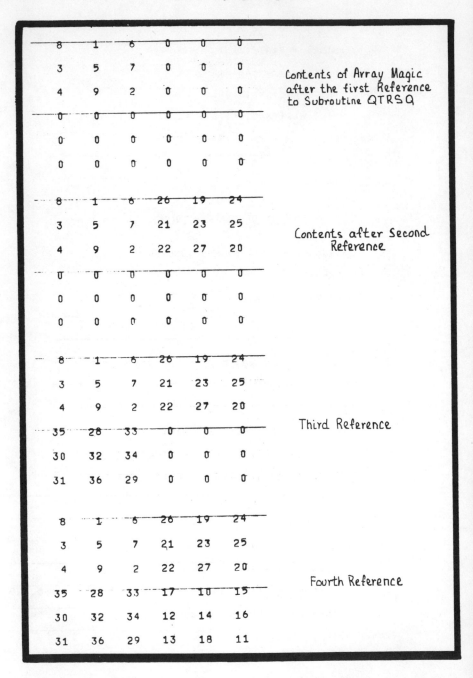

8	1	6	0	0	0	
3	5	7	0	0	0	Contents of Array Magic after the first Reference to Subroutine QTRSQ
4	9	2	0	0	0	
0	0	0	0	0	0	
0	0	0	0	0	0	
0	0	0	0	0	0	

8	1	6	26	19	24	
3	5	7	21	23	25	Contents after Second Reference
4	9	2	22	27	20	
0	0	0	0	0	0	
0	0	0	0	0	0	
0	0	0	0	0	0	

8	1	6	26	19	24	
3	5	7	21	23	25	
4	9	2	22	27	20	
35	28	33	0	0	0	Third Reference
30	32	34	0	0	0	
31	36	29	0	0	0	

8	1	6	26	19	24	
3	5	7	21	23	25	
4	9	2	22	27	20	
35	28	33	17	10	15	Fourth Reference
30	32	34	12	14	16	
31	36	29	13	18	11	

Figure 2-31. Contents of array MAGIC after the 6 by 6 magic square generating program references the subroutine QTRSQ.

```
      GO TO 10
C     KTR LESS THAN JSIZE/2--MOVE RIGHT AND JP
00050 KTR = KTR + 1
      I = I - 1
      J = J + 1
C     DOES NEW I AND J MOVE OUTSIDE SQUARE (TOP)
      IF (I - II + 1)70,60,70
C     YES-RESET ROW INDICATOR TO LAST ROW OF QUARTER SQUARE
00060 I = N
      GO TO 10
C     DOES NEW I AND J MOVE OUTSIDE SQUARE (RIGHT SIDE)
00070 IF(J - M) 10,10,80
C     YES - RESET COLUMN INDICATOR TO FIRST COLUMN OF QUARTER SQUARE
00080 J = JJ
      GO TO 10
00090 PRINT 100
00100 FORMAT( 6H0ERROR )
      END
```

Figure 2-31 illustrates the contents of Array MAGIC immediately after the subroutine QTRSQ returns program control back to the main program.

After returning from the subroutine for the fourth and last time, the main program transposes six cell locations and prints the 6 by 6 Magic Square shown in Figure 2-32.

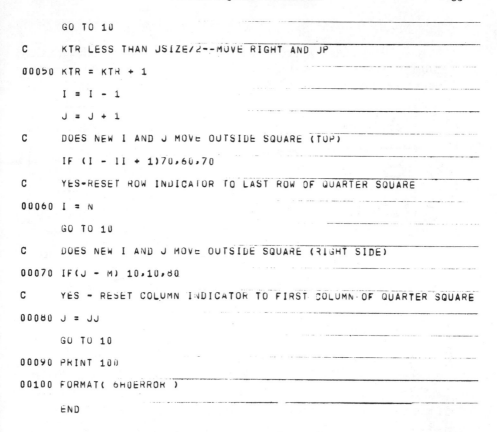

35	1	6	26	19	24
3	32	7	21	23	25
31	9	2	22	27	20
8	28	33	17	10	15
30	5	34	12	14	16
4	36	29	13	18	11

Figure 2-32. A magic square generated by the 6 by 6 magic square generating program.

2.6   Magic Squares Starting with Numbers Other Than One

Most of the magic squares in the previous sections started with the number 1. However, magic squares may be started with any number. Figure 2-33 illustrates a 3 by 3 Magic Square starting with 4.

11	4	9
6	8	10
7	12	5

Figure 2-33. A 3 by 3 magic square starting with 4.

107	100	105
102	104	106
103	108	101

Figure 2-34. A 3 by 3 magic square starting with 100.

The magic number of this square is 24 and is computed by the formula

$$\frac{\text{MAGIC}}{\text{NUMBER}} = \frac{n^3+n}{2} + n(p-1) = \frac{3^3+3}{2} + 3(4-1) = 24$$

where n is the order of the square and p is the starting number.

Another 3 by 3 Magic Square starting with 100 is shown in Figure 2-34. The magic number of this square is 312.

Figure 2-35 illustrates an Order 4 Square that starts with 3.

3	15	10	17
11	16	5	14
13	6	19	8
18	9	12	7

Figure 2-35. A 4 by 4 magic square starting with 3.

Magic Square Generator Program. The *Magic Square Generator Program* will generate an odd Magic Square starting with any number. The data input card to the program is illustrated in Figure 2-36 which describes the input information needed by the program.

After reading in the input card the program will print the heading:

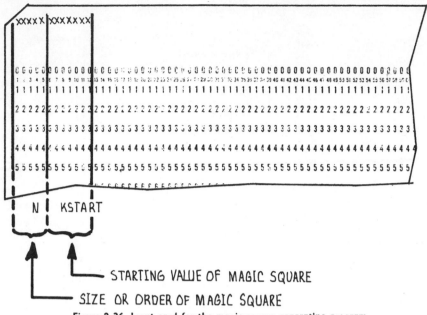

STARTING VALUE OF MAGIC SQUARE

SIZE OR ORDER OF MAGIC SQUARE

Figure 2-36. Input card for the magic square generating program.

MAGIC SQUARE N BY N STARTING WITH P

where N is the size of the magic square and P is the starting value.

The program sets the subscripts K and L to locate the middle cell in the first row of the square MAGODD. The starting value is stored in this location. The starting value is increased by 1 and a check is made to see if the program has stored N^2 values in Array MAGODD. If all values have been stored, the program will output the magic square. If the program is not through calculating, then another check is made to determine if KTR is an even multiple of N, and if so, KTR is reset to 1 and the row indicator K is advanced to the next row. If KTR is not a multiple of N, the KTR is advanced by 1 and the subscripts K and L are set to address the next cell of Array MAGODD which is to the *right* and *up*. If the new value of L indicates a cell location outside the right side of array MAGODD, then the column indicator L is reset to the first column of the array. If the new value of K is less than 1 then K is reset to N. The program then stores the correct number in the Array MAGODD and the program continues until KSTART exceeds the maximum value to be stored in the array. A flowchart of the program is shown in Figure 2-37 and a FORTRAN program follows.

The 15 by 15 Magic Square shown in Figure 2-38 was generated by this program. The starting value of this square is 7 and the magic number is

$$\text{MAGIC NUMBER} = \frac{15^3 + 15}{2} \; 15(7-1) = 1785$$

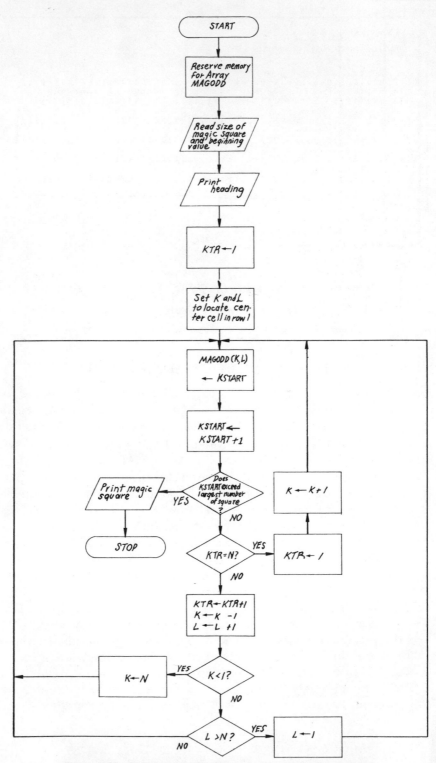

Figure 2-37. Flowchart of the magic square generating program.

```
C       MAGIC SQUARE GENERATOR ---
C       START WITH ANY NUMBER
        DIMENSION MAGODD(15,15)
C       READ MAGIC SQUARE SIZE AND STARTING NUMBER
        READ 10,N,KSTART
00010 FORMAT(I5,I7)
        PRINT 12,N,N,KSTART
00012 FORMAT(13H1MAGIC SQUARE   ,I3,1X,2HBY,I3,14H STARTING WITH ,I3 )
        MAX = N**2 + KSTART - 1
        KTR = 1
C       SET ROW AND COLUMN LOCATORS TO LOCATE CENTER
C       CELL IN ROW 1
        K = 1
        L = (N+1) / 2
C       PLACE NUMBER IN CELL OF MAGIC SQUARE
00015 MAGODD(K,L) = KSTART
        KSTART = KSTART + 1
C       DOES KSTART EXCEED LAST NUMBER OF SQUARE
        IF(KSTART - MAX) 30,30,20
00020 PRINT 25,((MAGODD(K,L),L=1,N),K=1,N)
00025 FORMAT(1H0 / 15I5)
        STOP
00030 IF(KTR - N) 50,40,40
C       RESET KTR AND SET K TO ONE ROW DOWN
00040 KTR = 1
        K = K + 1
        GO TO 15
C       ADD 1 TO KTR AND MOVE UP AND RIGHT
00050 KTR = KTR + 1
        K = K - 1
        L = L + 1
C       DID NEW K GENERATE LOCATION
C       OUTSIDE TOP OF SQUARE
        IF(K)70,60,70
C       YES -- SET K TO LAST ROW
```

```
00060 K = N
      GO TO 15
C     DID NEW L GENERATE LOCATION
C     OUTSIDE RIGHT SIDE OF SQUARE
00070 IF(L - N)15,15,80
C     YES -- RESET L TO FIRST COLUMN
00080 L = 1
      GO TO 15
      END
```

128	145	162	179	196	213	230	7	24	41	58	75	92	109	126
144	161	178	195	212	229	21	23	40	57	74	91	108	125	127
160	177	194	211	228	20	22	39	56	73	90	107	124	141	143
176	193	210	227	19	36	38	55	72	89	106	123	140	142	159
192	209	226	18	35	37	54	71	88	105	122	139	156	158	175
208	225	17	34	51	53	70	87	104	121	138	155	157	174	191
224	16	33	50	52	69	86	103	120	137	154	171	173	190	207
15	32	49	66	68	85	102	119	136	153	170	172	189	206	223
31	48	65	67	84	101	118	135	152	169	186	188	205	222	14
47	64	81	83	100	117	134	151	168	185	187	204	221	13	30
63	80	82	99	116	133	150	167	184	201	203	220	12	29	46
79	96	98	115	132	149	166	183	200	202	219	11	28	45	62
95	97	114	131	148	165	182	199	216	218	10	27	44	61	78
111	113	130	147	164	181	198	215	217	9	26	43	60	77	94
112	129	146	163	180	197	214	231	8	25	42	59	76	93	110

Figure 2-38. A 15 by 15 magic square starting with 7.

2.7 IXOHOXI Magic Square

The square of Figure 2-39 is composed entirely of numbers made up of the digits 1 and 8 and has additional properties beyond the normal requirement of all magic squares: that the sum of all rows, columns and both main diagonals be equal.

8818	1111	8188	1881		1118	8181	1888	8811
8181	1888	8811	1118		8888	1811	8118	1181
1811	8118	1181	8888		8111	1188	8881	1818
1188	8881	1818	8111		1881	8818	1111	8188

Figure 2-39. IXOHOXI magic square. Figure 2-40. Upside down IXOHOXI magic square.

Using the following position square

M_{11}	M_{12}	M_{13}	M_{14}
M_{21}	M_{22}	M_{23}	M_{24}
M_{31}	M_{32}	M_{33}	M_{34}
M_{41}	M_{42}	M_{43}	M_{44}

the additional arrangements are

$$\left.\begin{array}{l}
M_{33} + M_{34} + M_{43} + M_{44} \\
M_{11} + M_{14} + M_{41} + M_{44} \\
M_{21} + M_{31} + M_{24} + M_{34} \\
M_{13} + M_{24} + M_{31} + M_{42} \\
M_{12} + M_{13} + M_{42} + M_{43} \\
M_{22} + M_{23} + M_{32} + M_{33} \\
M_{21} + M_{12} + M_{43} + M_{34} \\
M_{11} + M_{12} + M_{21} + M_{22} \\
M_{13} + M_{14} + M_{23} + M_{24} \\
M_{21} + M_{22} + M_{31} + M_{32} \\
M_{23} + M_{24} + M_{33} + M_{34} \\
M_{31} + M_{32} + M_{41} + M_{42}
\end{array}\right\} = 19998$$

where 19998 is the *magic number*.

If the IXOHOXI * square is turned upside down, the reader will find that it is still a magic square. (See Figure 2-40.)

In fact not only are the rows, columns and diagonals equal, but all the number configurations of the original IXOHOXI square still add up to 19998 using the upside down square.

The mirror images of the IXOHOXI square and the upside down version of the square will have the same properties as do the original versions of the number arrangements. That is, all of the mentioned number arrangements apply to the mirror image squares as well as to the original and the upside down square. Mirror images of the IXOHOXI square are shown in Figure 2-41.

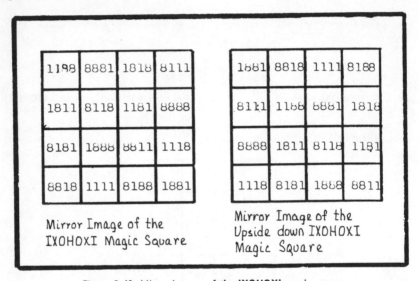

Figure 2-41. Mirror images of the IXOHOXI magic square.

IXOHOXI Magic Square Checking Program. The *IXOHOXI Magic Square Checking Program* inputs one of the following arrangements of the IXOHOXI Magic Square:

a. Normal Arrangement of IXOHOXI Magic Square
b. Upside Down IXOHOXI Magic Square
c. Mirror Image of the Normal Arrangement
d. Mirror Image of the Upside Down Square

The program determines the sum of each row, column, diagonal and group of consecutive 4-cell squares. If this sum equals 19998 (the magic number), an appropriate message is printed.

The flowchart of the IXOHOXI checking program is shown in Figure 2-42 and the FORTRAN program follows.

---

* Look at the name IXOHOXI in any way and it will always be the same: backwards, mirror image, upside down, etc.

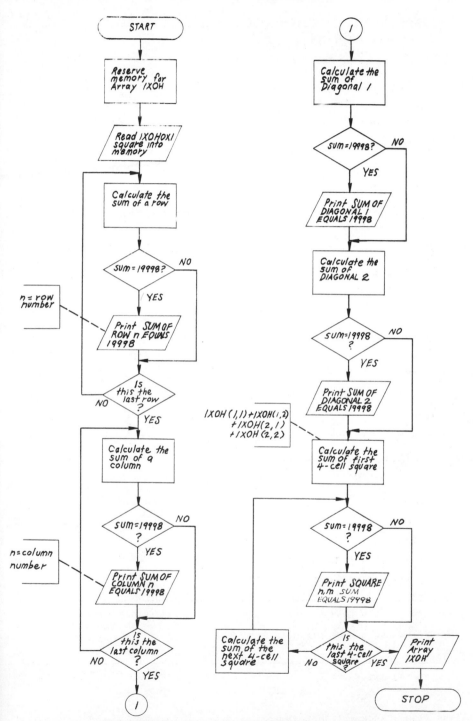

Figure 2-42. Flowchart of the IXOHOXI checking program.

```
C     IXOHOXI MAGIC SQUARE CHECKING PROGRAM
      DIMENSION IXOH(4,4)
      READ 10, ((IXOH(I,J),I=1,4),J=1,4)
00010 FORMAT (8I5)
C     DETERMINE SUM OF EACH ROW
      DO 50 I=1,4
      KSUM = 0
      DO 20 J=1,4
00020 KSUM = KSUM + IXOH(I,J)
      IF (KSUM - 19998) 50,30,50
00030 PRINT 40, I
00040 FORMAT (12H0 SUM OF ROW ,I3,29H EQUALS 19998 (MAGIC NUMBER)  / )
00050 CONTINUE
C     DETERMINE SUM OF EACH COLUMN
      DO 90 J=1,4
      KSUM = 0
      DO 60 I=1,4
00060 KSUM = KSUM + IXOH(I,J)
      IF(KSUM -19998) 90,70,90
00070 PRINT 80, J
00080 FORMAT (15H0 SUM OF COLUMN ,I3,28H EQUALS 19998 (MAGIC NUMBER) / )
00090 CONTINUE
C     DETERMINE SUM OF DIAGONAL 1
      KSUM = 0
      DO 100 I=1,4
      J=I
00100 KSUM = KSUM + IXOH(I,J)
      IF (KSUM - 19998) 130,110,130
00110 PRINT 120
00120 FORMAT (47H  SUM OF DIAGONAL 1 EQUALS 19998 (MAGIC NUMBER) ,/ )
C     DETERMINE SUM OF DIAGONAL 2
00130 KSUM =0
      DO 140 I=1,4
      J = 4 - I + 1
00140 KSUM = KSUM + IXOH(I,J)
```

```
        IF(KSUM - 19998) 170,150,170
00150 PRINT 160
00160 FORMAT (47H   SUM OF DIAGONAL 2 EQUALS 19998 (MAGIC NUMBER) ,/ )
C       DETERMINE IF OTHER SETS OF 4 CELLS EQUAL THE MAGIC NUMBER
00170 DO 220 L=1,3
        LL = L + 1
        DO 220 M = 1,3
        MM = M + 1
        KSUM = 0
        DO 190 I = L,LL
        DO 190 J = M,MM
00190 KSUM = KSUM + IXOH(I,J)
        IF(KSUM - 19998)220,200,220
00200 PRINT 210, M,L
00210 FORMAT (8H   SQUARE,I2,2H ,,I2,35H   SUM EQUALS 19998 (MAGIC NUMBER
     1) / )
00220 CONTINUE
        PRINT 230, ((IXOH(I,J),J=1,4),I=1,4)
00230 FORMAT (1H1,/,(4I8///))
        STOP
        END
```

The following problem serves to illustrate how the IXOHOXI checking program functions. Input to the program is the mirror image of the Upside Down IXOHOXI Magic Square. The input data cards are illustrated in Figure 2-43.

The program first determines the sum of Row 1 and prints the message:

SUM OF ROW 1 EQUALS 19998 (MAGIC NUMBER)

The next row is then summed and another similar message is printed. The program continues this checking function and continues to print out diagnostic messages. After the program checks all rows, columns and main diagonals, the input array of numbers is printed as shown in Figure 2-44.

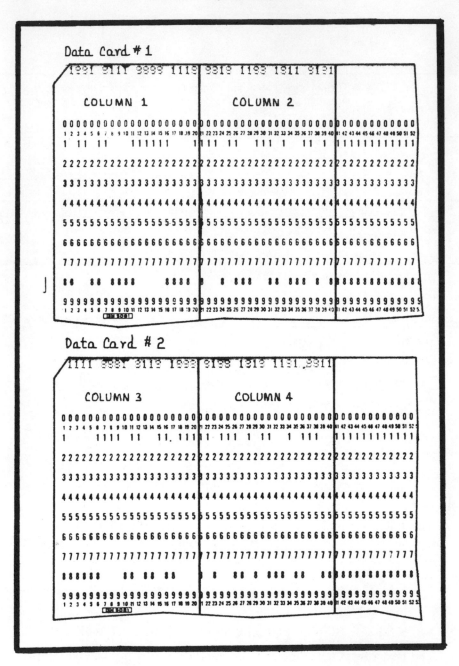

Figure 2-43. Input data cards for the IXOHOXI magic square checking program.

```
SUM OF ROW   1 EQUALS 19998 (MAGIC NUMBER)

SUM OF ROW  :2 EQUALS 19998 (MAGIC NUMBER)

SUM OF ROW   3 EQUALS 19998 (MAGIC NUMBER)

SUM OF ROW   4 EQUALS 19998 (MAGIC NUMBER)

SUM OF COLUMN   1 EQUALS 19998 (MAGIC NUMBER)

SUM OF COLUMN   2 EQUALS 19998 (MAGIC NUMBER)

SUM OF COLUMN   3 EQUALS 19998 (MAGIC NUMBER)

SUM OF COLUMN   4 EQUALS 19998 (MAGIC NUMBER)

SUM OF DIAGONAL 1 EQUALS 19998 (MAGIC NUMBER)

SUM OF DIAGONAL 2 EQUALS 19998 (MAGIC NUMBER)

SQUARE 1 , 1   SUM EQUALS 19998 (MAGIC NUMBER)

SQUARE 2 , 1   SUM EQUALS 19998 (MAGIC NUMBER)

SQUARE 3 , 1   SUM EQUALS 19998 (MAGIC NUMBER)

SQUARE 1 , 2   SUM EQUALS 19998 (MAGIC NUMBER)

SQUARE 2 , 2   SUM EQUALS 19998 (MAGIC NUMBER)

SQUARE 3 , 2   SUM EQUALS 19998 (MAGIC NUMBER)

SQUARE 1 , 3   SUM EQUALS 19998 (MAGIC NUMBER)

SQUARE 2 , 3   SUM EQUALS 19998 (MAGIC NUMBER)

SQUARE 3 , 3   SUM EQUALS 19998 (MAGIC NUMBER)
```

1881	8818	1111	8188
8111	1188	8981	1818
8888	1811	8118	1181
1118	8181	1888	8811

Figure 2-44. Output of the IXOHOXI magic square checking program.

2.8   Multiplication Magic Squares

Figure 2-45 illustrates a 3 by 3 Multiplication Magic Square. The magic number is 216 and is obtained by multiplying together the three numbers in any column, row or diagonal.

12	1	18
9	6	4
2	36	3

Figure 2–45. A 3 by 3 multiplication magic square.

A method based on the De la Loubere odd order constructing method may be used to generate multiplication magic squares of odd order. The construction of a 5 by 5 Multiplication Square is used to illustrate the method.

1. Place the number 1 in the center cell of the first row in a blank 5 by 5 square.

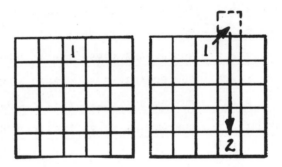

2. Move in an oblique direction, one square above and to the right. This movement results in leaving the top of the box. It is necessary to place the next number in a cell at the bottom of the column in which you wanted to place the number. The number to place at this location is twice the last number or 2.

3. Now move diagonally to the right again and put the number twice the last or 4 in the next cell location.

4. If you continue diagonally to the right you leave the cell on the right side. When this occurs, you must go to the extreme left of the row in which you wanted to place the next number. After crossing over to the left side of the square, place the number twice that of the last into the appropriate cell.

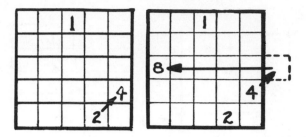

5. Now, again, go up diagonally to the right and place the next number. This number is determined by doubling the last number.

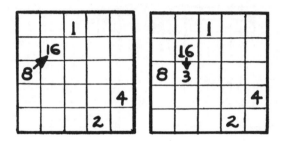

This completes the first group of 5 numbers of the 5 by 5 square. The next group of 5 numbers starts with 3, the next group 9, the next group 27 and the last group 81. The reader should note that the starting numbers are all powers of 3:

$$3^0 = 1 \qquad 3^2 = 9 \qquad 3^4 = 81$$
$$3^1 = 3 \qquad 3^3 = 27$$

6. Since this is a 5 by 5 square, you must move down one cell to place the next group of 5 numbers. In the case of a 3 by 3 square, you would move down when you reached a group of 3. The number placed in this cell is the starting number of the second group of 5 numbers.

7. Move up diagonally to the right and place a number into each cell you enter. If you leave the top of the box, move to the bottom of the column where you wanted to place the number. If you move outside the box on the right side, move across to the opposite side. The next number is always determined by doubling the last number. When you finish the second group of five numbers, the square should appear as shown at the left on the top of the next page.

8. The following square would result after the third group of 5 numbers have been placed. The starting number is 9.

9. The fourth group of 5 numbers would be placed in the following square. The starting number is 27.

10. When the 25th number is placed in the cell opposite the starting cell, the final square will appear as shown.

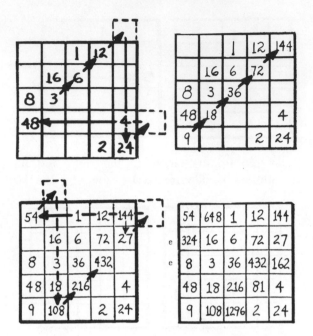

Multiplication Magic Square Generating Program. This program will generate a 5 by 5 Multiplication Magic Square by a method similar to that of De la Loubere. The program reserves memory for the magic square in Array MAGIC. Starting values for NGROUP, KTR and NUM are 1. NGROUP is a variable that contains the starting value of each group of five numbers. KTR is a counter that is used to determine multiples of 5. NUM is a variable that will vary from the starting number 1 to the 25th number 1296. NUM contains the value that is stored in each cell of Array MAGIC. The subscripts, I and J, are first to specify the center cell of row 1 of Array MAGIC. After the value of NUM is stored in cell MAGIC (I,J), a new value of NUM is computed, twice that of the old value. This computed value is compared against the largest number in the square, 1296. If the value of NUM exceeds 1296, then the program prints the heading

MULTIPLICATION MAGIC SQUARE 5 BY 5

and the contents of Array MAGIC. If the value of NUM did not exceed 1296, the program continues and a check is made to see if the KTR is a multiple of N or 5. If it is, then KTR is reset to 1, the *row* subscript is set to specify the next *row*, a starting value for the next group of five numbers is computed (NUM = 3*NGROUP), and the variable NGROUP is set to this computed value of NUM. If KTR is not a multiple of N, then KTR is increased by 1 and the subscripts I and J are updated to specify the next cell to the *right* and *up*. If the new value of the row subscript I is less than 1, which indicates a cell outside the *top* of the

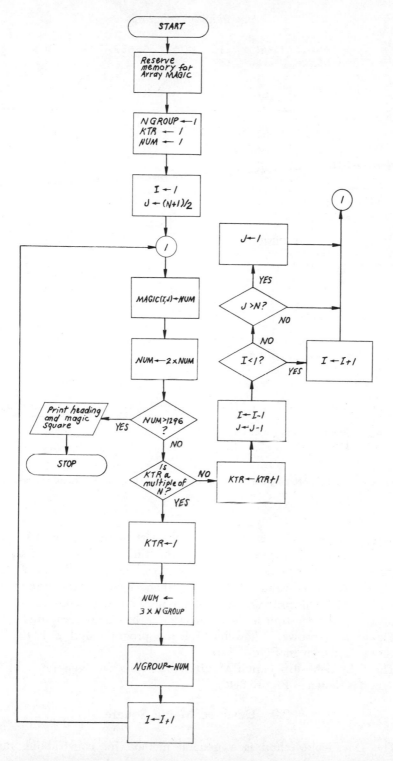

Figure 2-46. Flowchart of the multiplication magic square generating program.

```
C         MULTIPLICATION MAGIC SQUARE GENERATOR
          DIMENSION MAGIC(5,5)
          N = 5
          NGROUP = 1
          KTR = 1
          NUM = 1
C         LOCATE FIRST CELL OF ARRAY MAGIC
          I = 1
          J = (N+1) / 2
   10 MAGIC(I,J) = NUM
          NUM = 2 * NUM
          IF(NUM - 1296)11,11,55
   11 IF(KTR - N)30,20,20
   20 KTR = 1
          I = I + 1
          NUM = 3 * NGROUP
          NGROUP = NUM
          GO TO 10
   30 KTR = KTR + 1
          I = I - 1
          J = J + 1
          IF(I)40,35,40
   35 I = N
          GO TO 10
   40 IF(J-N)10,10,50
   50 J = 1
          GO TO 10
   55 PRINT 60,N,N
   60 FORMAT(29H1 MULTIPLICATION MAGIC SQUARE,I4,3H BY,I4)
          PRINT 65,((MAGIC(I,J),J=1,N),I=1,N)
   65 FORMAT(1H0 /////,5I10)
          STOP
          END
```

Figure 2-47. FORTRAN program of the multiplication magic square generating program.

square, I is set to address the cell in the last *row* of Array MAGIC. If the new value of J is greater than N, which indicated a cell outside the *right* side of the square, J is set to address the first column of Array MAGIC. The new value of NUM is then stored in the new cell of MAGIC (I,J). This process will continue until NUM exceeds 1296. When this occurs, the multiplication magic square has been computed.

Figure 2-46 shows a flowchart of the program and a FORTRAN program is shown in Figure 2-47.

The 5 by 5 Multiplication Magic Square that was generated by this program is shown in Figure 2-48.

2.9 Devedec Magic Square

The Devedec Method is a general method for constructing magic squares of any order.

MULTIPLICATION MAGIC
SQUARE 5 BY 5

54	648	1	12	144
324	16	6	72	27
8	3	36	432	162
48	18	216	81	4
9	108	1296	2	24

Figure 2-48. A 5 by 5 magic square generated by the multiplication magic square generating program.

To use this method one generates an even ordered square by substituting numbers from a *basic square* into a predetermined *coded square*. The numbers are placed using the following set of rules:

1. If a 1 appears in the *coded square*, then replace 1 with the number from the *basic square* that occupies the same position.

2. If a 2 appears in the *coded square*, then replace 2 with the number from the *basic square* that occupies the position symmetric to this position with respect to the center.

3. If a 3 appears in the *coded square*, then replace 3 with the number from the *basic square* that occupies the position symmetric to this position with respect to the horizontal axis.

4. If a 4 appears in the *coded square*, then replace 4 with the number from the *basic square* that occupies the position symmetric to this position with respect to the vertical axis.

The *basic square* is composed of numbers written in the order specified by Figure 2-49.

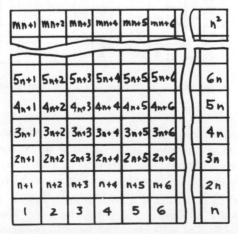

Figure 2-49. Basic square for the Devedec method.

1	3	4	3	4	3	3	1	3	4	3	4	3	1
4	1	3	4	3	4	2	2	4	3	4	3	1	2
3	4	1	3	4	3	3	1	3	4	3	1	2	1
4	3	4	1	3	4	2	2	4	3	1	2	1	2
3	4	3	4	1	3	3	1	3	1	2	1	2	1
4	3	4	3	4	1	3	3	1	2	1	2	1	2
4	2	4	2	4	4	2	2	2	4	2	4	2	4
1	2	1	2	1	4	2	2	2	1	2	1	2	1
4	3	4	3	4	1	2	2	1	2	1	2	1	2
3	4	3	4	1	2	3	1	2	1	2	1	2	1
4	3	4	1	2	1	2	2	1	2	1	2	1	2
3	4	1	2	1	2	3	1	2	1	2	1	2	1
4	1	2	1	2	1	2	2	1	2	1	2	1	2
1	2	1	2	1	2	3	1	2	1	2	1	2	1

Figure 2-50. Coded square for constructing magic squares of order 2(2n+1) by the Devedec method.

1	3	4	3	4	2	2	4	3	4	3	1
4	1	3	4	3	3	1	3	4	3	1	2
3	4	1	3	4	2	2	4	3	1	2	1
4	3	4	1	3	3	1	3	1	2	1	2
3	4	3	4	1	2	2	1	2	1	2	1
2	4	2	4	2	1	1	2	4	2	4	2
2	1	2	1	2	1	1	2	1	2	1	2
3	4	3	4	1	2	2	1	2	1	2	1
4	3	4	1	2	3	1	2	1	2	1	2
3	4	1	2	1	2	2	1	2	1	2	1
4	1	2	1	2	3	1	2	1	2	1	2
1	2	1	2	1	2	2	1	2	1	2	1

Figure 2-51. Coded square for constructing magic squares of order 4n by the Devedec method.

This method uses two *coded squares*. The *coded square* of Figure 2-50 is used to generate singly-even order squares, up to and including 14 by 14 Magic Squares. Figure 2-51 illustrates a *coded* square that can be used to generate doubly-even order squares up to 12 by 12 squares.

If one wanted to construct a 4 by 4 Magic Square, the center square of the *coded square* in Figure 2-51 would be used. The center square of the *coded square* Figure 2-50 would be used to generate a 6 by 6 square. The middle square of Figure 2-51 would be used to construct an 8 by 8

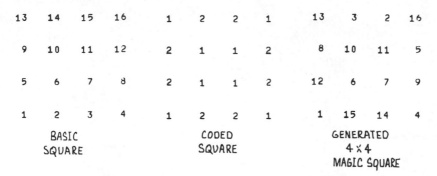

13	14	15	16	1	2	2	1	13	3	2	16
9	10	11	12	2	1	1	2	8	10	11	5
5	6	7	8	2	1	1	2	12	6	7	9
1	2	3	4	1	2	2	1	1	15	14	4

BASIC SQUARE   CODED SQUARE   GENERATED 4 x 4 MAGIC SQUARE

Figure 2-52. Example showing the generation of a 4 by 4 magic square by the Devedec method.

31	32	33	34	35	36
25	26	27	28	29	30
19	20	21	22	23	24
13	14	15	16	17	18
7	8	9	10	11	12
1	2	3	4	5	6

1	3	3	1	3	1
4	1	3	3	1	2
4	4	2	2	2	4
1	4	2	2	2	1
4	1	2	2	1	2
1	2	2	1	2	1

31	2	3	34	5	36
30	26	9	10	29	7
24	23	16	15	14	19
13	17	22	21	20	18
12	8	28	27	11	25
1	35	34	4	32	6

BASIC SQUARE   CODED SQUARE   GENERATED 6 BY 6 MAGIC SQUARE

Figure 2-53. Example showing the generation of a 6 by 6 magic square by the Devedec Method.

square. The middle square of Figure 2-50 would be used to construct a 10 by 10 square. A 12 by 12 would be constructed using the entire square shown in Figure 2-51. A 14 by 14 square could be constructed using the coded square of Figure 2-50. For example, the construction of a 4 by 4 Magic Square is shown in Figure 2-52.

Figure 2-53 illustrates the construction of an Order 6 Magic Square. The reader should note that both Order *4n* and Order *2(2n+1)* Magic Squares can be constructed by this method simply by referencing a different *coded square*.

Devedec Magic Square Generating Program. This program can be used to generate even order magic squares up to and including order 12 squares. Only minor revisions would be required to cause the program to generate larger squares: DIMENSION larger arrays, change FORMAT statements 20 and 130, and prepare the data cards for the larger *coded square*.

The program reserves computer memory for the *basic square* (array IBASIC), the *coded square* (array ICODE) and the area where the magic square will be generated (array IGEN). The program first reads a card which contains the *order* of the square to be generated (N) and

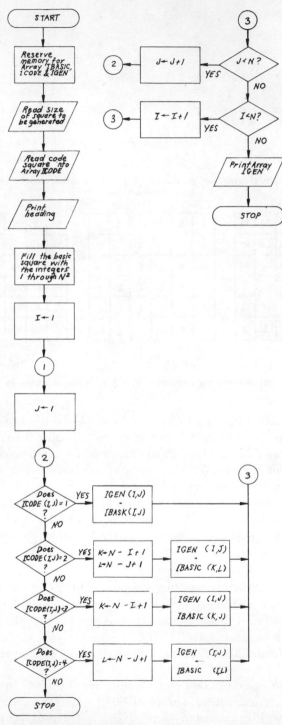

Figure 2-54. Flowchart of the Devedec magic square generating program.

then reads the *coded square* data cards into array ICODE. The program then prints the heading

MAGIC SQUARE n BY n

where n is the order of the square to be generated. The program then fills the *basic square* (IBASIC) with the numbers 1 through N^2 starting in the cell in the lower left corner of the square. The program then cycles through each location of Array ICODE determining what value of Array IBASIC to store in Array IGEN. The storing logic used in the program is identical to the four basic replacement rules of the Devedec method. After the N^2 location of Array IGEN has been determined, the program prints the entire array. Figure 2-54 illustrates a flowchart of the program.

The following FORTRAN program is written to generate a 12 by 12 Magic Square. The program reads a data card with the value 12 punched in columns 4 and 5, and the 12 data cards which contain the *coded square*. These cards are illustrated in Figure 2-55.

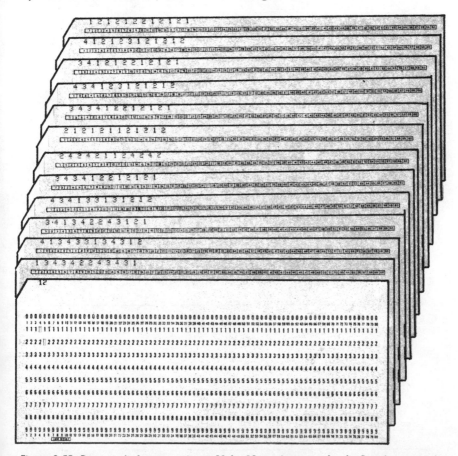

Figure 2-55. Data cards for generating a 12 by 12 magic square by the Devedec generating program.

```
C      EVEN MAGIC SQUARE GENERATING PROGRAM (DEVEDEC METHOD)
       DIMENSION ICODE(12,12),IBASIC(12,12),IGEN(12,12)
C      READ SIZE OF SQUARE TO BE GENERATED
       READ 10,N
00010 FORMAT(I5)
C      READ CODED SQUARE
       READ 20,((ICODE(I,J),J=1,N),I=1,N)
00020 FORMAT(12I2)
       PRINT 30,N,N
00030 FORMAT(13H1MAGIC SQUARE ,I4,2X,2HBY,I4)
C      FILL BASIC SQUARE (IBASIC) WITH THE VALUES 1 THRU N**2
       NUM = 1
       DO 40 I = 1,N
       DO 40 J = 1,N
       K = N - I + 1
       L = N - J + 1
       IBASIC(K,L) = NUM
00040 NUM = NUM + 1
C      CHECK VALUES OF CODED SQUARE
       DO 120 I = 1,N
       DO 120 J = 1,N
       IF(ICODE(I,J) - 1) 60,50,60
C      A 1 APPEARS IN THIS CELL OF CODED SQUARE
00050 IGEN(I,J) = IBASIC(I,J)
       GO TO 120
00060 IF(ICODE(I,J) - 2)80,70,80
C      A 2 APPEARS IN THIS CELL OF CODED SQUARE
00070 K = N - I + 1
       L = N - J + 1
       IGEN(I,J) = IBASIC(K,L)
       GO TO 120
00080 IF(ICODE(I,J) - 3)100,90,100
C      A 3 APPEARS IN THIS CELL OF CODED SQUARE
00090 K = N - I + 1
       IGEN(I,J) = IBASIC(K,J)
       GO TO 120
```

```
00100 IF(ICODE(I,J) - 4)140,110,140
C      A 4 APPEARS IN THIS CELL OF CODED SQUARE
00110 L = N - J + 1
      IGEN(I,J) = IBASIC(I,L)
00120 CONTINUE
      PRINT 130,((IGEN(I,J),J=1,N),I=1,N)
00130 FORMAT(1H0,12I5 / )
00140 STOP
```

Output from the program is the 12 by 12 Magic Square shown in Figure 2-56.

144	11	135	9	137	6	7	140	4	142	2	133
121	131	22	124	20	19	126	17	129	15	122	24
36	110	113	33	113	30	31	116	28	111	35	109
97	47	99	105	44	43	102	41	100	46	98	48
60	86	58	88	92	54	55	89	57	87	59	85
61	74	63	76	65	79	78	68	81	70	83	72
73	71	75	69	77	67	66	80	64	82	62	84
96	50	94	52	56	90	91	53	93	51	95	49
37	107	39	45	101	103	42	104	40	106	38	108
120	26	34	112	32	114	115	29	117	27	119	25
13	23	123	21	125	127	18	128	16	130	14	132
12	134	10	136	8	138	139	5	141	3	143	1

Figure 2-56. A 12 by 12 magic square generated by the Devedec generating program.

2.10   Domino Magic Squares

A 3 by 3 Magic Square constructed of dominoes appears in Figure 2-57.

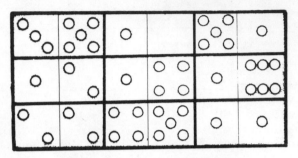

Figure 2-57. A 3 by 3 domino magic square.

The magic number of this square is 15 and the sum of each row, column and diagonal is determined as follows:

ROW 1 $\boxed{3\,|\,5}$ + $\boxed{1\,|\,\emptyset}$ + $\boxed{5\,|\,1}$ = 8 + 1 + 6 = 15

ROW 2 $\boxed{1\,|\,2}$ + $\boxed{1\,|\,4}$ + $\boxed{1\,|\,6}$ = 3 + 5 + 7 = 15

ROW 3 $\boxed{2\,|\,2}$ + $\boxed{4\,|\,5}$ + $\boxed{1\,|\,1}$ = 4 + 9 + 2 = 15

COLUMN 1 $\boxed{3\,|\,5}$ + $\boxed{1\,|\,2}$ + $\boxed{2\,|\,2}$ = 8 + 3 + 4 = 15

COLUMN 2 $\boxed{1\,|\,\emptyset}$ + $\boxed{1\,|\,4}$ + $\boxed{4\,|\,5}$ = 1 + 5 + 9 = 15

COLUMN 3 $\boxed{5\,|\,1}$ + $\boxed{1\,|\,6}$ + $\boxed{1\,|\,1}$ = 6 + 7 + 2 = 15

DIAGONAL 1 $\boxed{3\,|\,5}$ + $\boxed{1\,|\,4}$ + $\boxed{1\,|\,1}$ = 8 + 5 + 2 = 15

DIAGONAL 2 $\boxed{5\,|\,1}$ + $\boxed{1\,|\,4}$ + $\boxed{2\,|\,2}$ = 6 + 5 + 4 = 15

The Domino Magic Square of Figure 2-57 was constructed using the De la Loubere method for constructing odd order magic squares.

1. Start by placing domino 1 (domino having dots that sum to 1, i.e., $\boxed{\,|\,\bullet}$) in the center cell of row 1. Move diagonally to the right and if you leave the box, go to the bottom of the column in which you wanted to place the domino. Put domino 2 (domino having dots that sum to 2, i.e., $\boxed{\bullet\,|\,\bullet}$, $\boxed{\,|\,\colon}$) in this box. Now again move diagonally to the right and if you leave the box, go to the left end of the row in which you wanted to place the domino. Put domino 3 (domino having dots that sum to 3, i.e., $\boxed{\bullet\,|\,\colon}$, $\boxed{\,|\,\therefore}$) in this location. This completes a group of 3 dominoes (on left at top of the next page.)

2. Since this is a 3 by 3 square it is necessary to move down one box

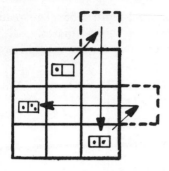
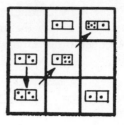

to generate the next set of 3 dominoes. Thus dominoes 4, 5 and 6 are the next set of dominoes and are placed as shown at right above.

3. Domino 7 is placed under domino 6 because 6 is a multiple of 3. Dominoes 8 and 9 are placed using the above rules and the completed Domino Magic Square will appear as shown in Figure 2-57.

Table 2-3 may prove helpful in the construction of Domino Magic Squares.

Table 2-3. A 6 by 6 Domino Set and Magic Square Values.

MAGIC SQUARE VALUE	DOMINO VALUE	MAGIC SQUARE VALUE	DOMINO VALUE
0		7	
1		8	
2		9	
3		10	
4		11	
5		12	
6			

Other 3 by 3 Magic Squares are shown in Figure 2-58. The reader should determine the magic number of each of these squares.

An even order Domino Magic Square appears in Figure 2-59. In order

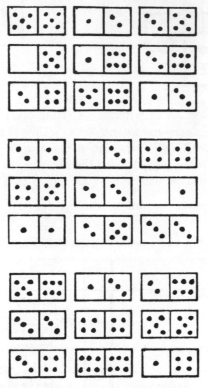

Figure 2-58. Three 3 by 3 domino magic squares.

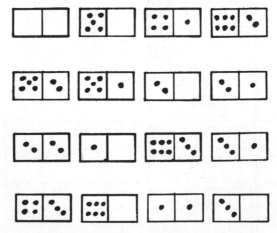

Figure 2-59. A 4 by 4 domino magic square.

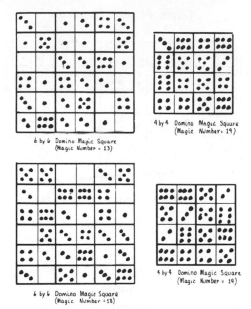

6 by 6 Domino Magic Square
(Magic Number = 13)

4 by 4 Domino Magic Square
(Magic Number = 19)

6 by 6 Domino Magic Square
(Magic Number = 18)

4 by 4 Domino Magic Square
(Magic Number = 19)

Figure 2-60. Domino magic squares.

to generate this square, it was necessary to use more than one domino having the same sum. For example, in COLUMN 2, both ⬚6⬚ and ⬚5⬚1⬚ add up to 6, and in ROW 3, both ⬚3⬚1⬚ and ⬚2⬚2⬚ add up to 4. One will find that that is necessary when generating Domino Magic Squares of Order 4 or larger. The magic number of this square is 18.

Other Domino Magic Squares are shown in Figures 2-60 and 2-61. Note that the 6 by 6 squares and two of the 4 by 4 squares are slightly

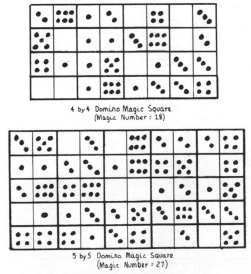

4 by 4 Domino Magic Square
(Magic Number = 18)

5 by 5 Domino Magic Square
(Magic Number = 27)

Figure 2-61. Domino magic squares.

5 by 5 Domino Magic Square
(Magic Number: 27)

Figure 2-61. Domino magic squares (cont'd).

different in that fewer dominoes are used in the construction of the squares.

3 By 3 Domino Magic Square Generating Program. This program uses the De la Loubere generating procedure. It will generate a 3 by 3 Domino Magic Square. The program reserves memory for two arrays: Array IDOM and Array MAGIC. The program then reads nine data cards which contain the values of the dominoes. These cards are illustrated in Figure 2-62.

The program then generates a magic square placing the appropriate domino values in Array MAGIC. When the last domino value is stored, the program prints the square. Figure 2-63 illustrates a flowchart of the program. A FORTRAN program is shown below.

```
C       DOMINO MAGIC SQUARE GENERATOR

        DIMENSION IDOM(9),MAGIC(3,3)

C       READ DOMINO VALUES INTO ARRAY IDOM

        READ 10,(IDOM(K),K=1,9)

00010   FORMAT(I5)

        KTR = 1

        NUM = 1

        I=1

        N=3

        J=(N+1)/2

C       PUT FIRST DOMINO VALUE IN CENTER CELL OF FIRST ROW

00020   MAGIC(I,J) = IDOM(NUM)

        NUM = NUM + 1

        IF(NUM - N**2)30,30,90

00030   IF(KTR - N)50,40,40

C       RESET KTR TO 1 AND SET I TO NEXT ROW
```

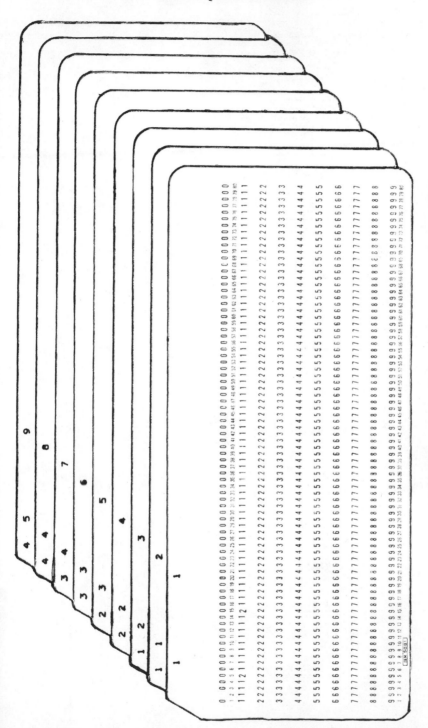

Figure 2-62. Input data cards for the 3 by 3 domino magic square generating program.

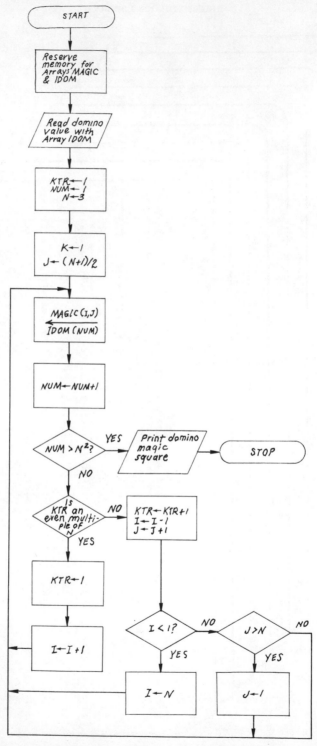

Figure 2-63. Flowchart of the 3 by 3 domino magic square generating program.

```
00040 KTR = 1

      I = I + 1

      GO TO 20

C     MOVE TO THE RIGHT AND UP AND INCREASE KTR

00050 KTR = KTR + 1

      I = I - 1

      J = J + 1

      IF(I)70,60,70

C     MOVE WENT OUT OF TOP OF SQUARE

00060 I = N

      GO TO 20

00070 IF(J-N)20,20,80

C     RESET COLUMN INDEX TO 1

00080 J = 1

      GO TO 20

C     PRINT HEADING AND DOMINO MAGIC SQUARE

00090 PRINT 100,N,N

00100 FORMAT(1H1,3X,I2,3H BY,I2,21H   DOMINO MAGIC SQUARE ,/)

      PRINT 110,((MAGIC(I,J),J=1,N),I=1,N)

00110 FORMAT(1H0,/,3(4X,I5,3X,I5,3X,I5,//)))

      STOP

      END
```

Output from the previous FORTRAN program is the 3 by 3 Domino Magic Square of Figure 2-64.

4004	1	3003
1002	2003	3004
2002	4005	1001

Figure 2-64. A 3 by 3 domino magic square printed by the domino magic square generating program.

4 By 4 Domino Magic Square Generating Program. The 4 by 4 Domino Magic Square shown in Figure 2-65 was generated by a modified version of the Devedec Method. A flowchart of this program is shown in Figure 2-66.

8008	1001	1002	6007
2003	5006	5005	4004
4005	3004	3003	6006
2002	7007	7008	1

Figure 2-65. A square generated by the 4 by 4 domino magic square generating program.

The FORTRAN program that is shown below first reads sixteen data cards into Array NUM. Each data card contains a domino value punched in columns 1, 2, 3, 4 and 5. A 9 by 9 set of dominoes was used for establishing the domino values for this program. The program clears all cell locations of Array MAG and then stores 77 in each cell that lies on a diagonal. The domino values are stored in Array MAG in a similar manner as were the 16 numbers of 4 by 4 Magic Square of Section 2.5.

```
C       4X4 DOMINO MAGIC SQUARE GENERATING PROGRAM
C       PROGRAM USES A 9 BY 9 SET OF DOMINOES
        DIMENSION NUM(16),MAG(4,4)
C       READ DOMINO VALUES INTO ARRAY NUM
        READ 5,(NUM(K),K=1,16)
00005 FORMAT(I5)
C       PLACE ZEROS IN ALL CELLS OF ARRAY MAG
        DO 10 I=1,4
        DO 10 J=1,4
00010 MAG(I,J)=0
C       STORE 77 IN EACH CELL OF DIAGONALS 1 AND 2
        DO 20 I=1,4
        J = I
00020 MAG(I,J) = 77
        DO 30 I=1,4
        J = 5 - I
00030 MAG(I,J) = 77
C       SET COUNTER TO 1
        KTR = 1
```

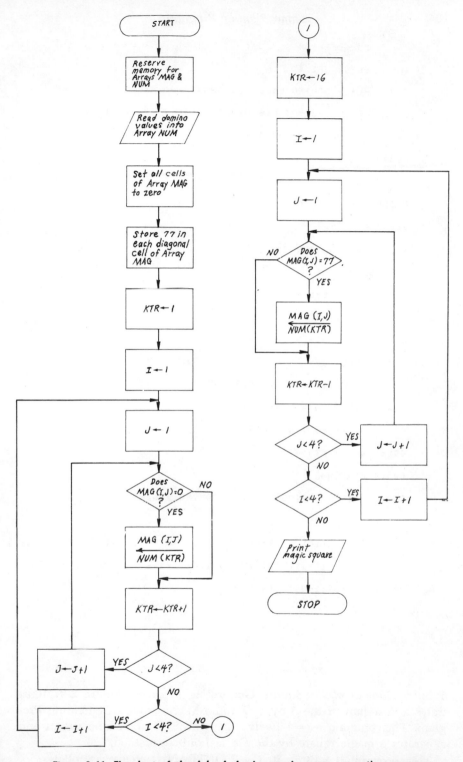

Figure 2-66. Flowchart of the 4 by 4 domino magic square generating program.

```
        DO 60 I=1,4

        DO 60 J=1,4

C       PLACE DOMINO VALUE IN ARRAY MAG IF

C       CELL DOES NOT LIE ON DIAGONAL (NOT EQUAL 77)

        IF (MAG(I,J))50,40,50

00040   MAG(I,J) = NUM(KTR)

C       INCREASE INDEX KTR

00050   KTR = KTR + 1

00060   CONTINUE

C       SET COUNTER TO 16

        KTR = 16

        DO 90 I=1,4

        DO 90 J=1,4

C       PLACE DOMINO VALUE IN ARRAY MAG IF

C       CELL LIES ON A DIAGONAL (EQUAL 77)

        IF(MAG(I,J)-77) 80,70,80

00070   MAG(I,J) = NUM(KTR)

C       DECREASE INDEX KTR

00080   KTR = KTR - 1

00090   CONTINUE

C       PRINT HEADING AND 4X4 DOMINO SQUARE

        PRINT 100

00100   FORMAT(1H1,6X,26H4 BY 4 DOMINO MAGIC SQUARE )

        PRINT 110,((MAG(I,J),J=1,4),I=1,4)

00110   FORMAT(1H0,/,4(4I8,//))

        STOP

        END
```

5 By 5 Domino Magic Square Generating Program. Figure 2-67 illustrates a flowchart of the 5 by 5 Domino Magic Square Generating Program. This program uses a twelve by twelve set of domino values and generates a magic square by the De la Loubere method.

This program is written in the FORTRAN language as follows:

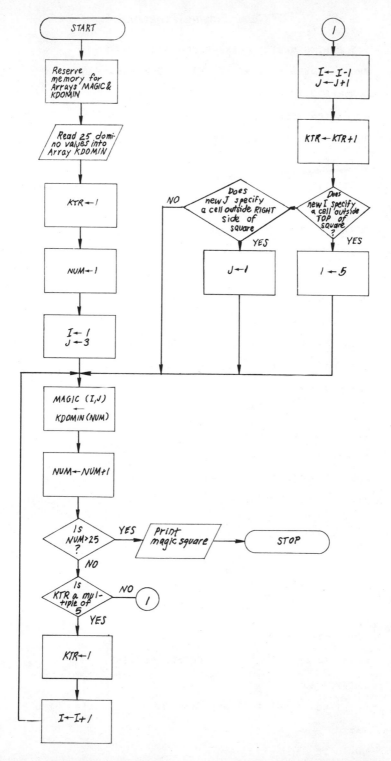

Figure 2-67. Flowchart of the 5 by 5 domino magic square generating program.

```
C     5X5 DOMINO MAGIC SQUARE CALCULATOR
C     PROGRAM USES A 12 BY 12 SET OF DOMINOES
C     NUMBERS 0 THRU 24
      DIMENSION KDOMIN(25),MAGIC(5,5)
C     READ 25 DOMINO VALUES INTO ARRAY KDOMIN
      READ 10,(KDOMIN(J),J=1,25)
00010 FORMAT(I5)
      KTR=1
      NUM=1
      I=1
      J=3
C     LOCATE CENTER CELL OF ROW 1 AND PLACE FIRST DOMINO VALUE
00020 MAGIC(I,J) = KDOMIN(NUM)
C     INCREMENT INDEX AND CHECK TO SEE IF IT NOW
C     EXCEEDS THE MAXIMUN NUMBER OF A 5X5 SQJARE
      NUM = NUM + 1
      IF(NUM - 25)30,30,90
C     IS KTR A MULTIPLE OF 5
00030 IF(KTR - 5)50,40,40
C     RESET KTR TO 1 IF YES AND SET I TO NEXT ROW
00040 KTR = 1
      I = I + 1
      GO TO 20
C     ADJUST INDEXES I AND J AND INCREASE KTR
00050 I = I - 1
      J = J + 1
      KTR = KTR + 1
C     CHECK TO SEE IF I IS OUT CF SQUARE (I=0)
      IF(I)70,60,70
C     RESET ROW INDEX TO 5 AND JUMP TO PLACE NEXT DOMINO VALUE
00060 I=5
      GO TO 20
```

```
C       CHECK TO SEE IF J IS OUT OF SQUARE(J=6)
00070 IF(J-5)20,20,80
C       RESET COLUMN INDEX TO 1 AND JUMP TO PLACE NEXT DOMINO VALUE
00080 J=1
        GO TO 20
C       PRINT HEADING AND 5X5 DOMINO MAGIC SQUARE
00090 PRINT 100
00100 FORMAT(1H1,9X,26H5 BY 5 DOMINO MAGIC SQUARE )
        PRINT 110,((MAGIC(I,J),J=1,5),I=1,5)
00110 FORMAT(1H0,/,5(2X,I5,3X,I5,3X,I5,3X,I5,3X,I5,//))
        STOP
        END
```

The 5 by 5 square of Figure 2-68 was printed by this program.

8008	11012	0	3004	7007
11011	2002	3003	6007	7008
1002	2003	6006	9010	10011
4005	5006	9009	10010	1001
5005	8009	12012	1	4004

Figure 2-68. A square generated by the 5 by 5 domino magic square generating program.

CHAPTER 3

Prime Numbers

3.1 What Is A Prime Number?

A *prime number* OR simply a *prime* is an integer that can be evenly divided only by the number 1 and the number itself. Prime numbers are the building blocks that form all other integers. The first 10 prime and 15 non-prime numbers are shown in Figure 3-1. The reader should note that each of the integers is either a prime number or is the sum of two or more primes.

1 — prime
2 — prime
3 — prime
4 — sum of 2+2
5 — prime
6 — sum of 2+2+2; 2+3+1; 5+1
7 — prime
8 — sum of 2+2+2+2; 5+3; 2+2+3+1
9 — sum of 3+3+3; 5+3+1; 7+2

.
.

.
25 — sum of 5+5+5+5+5; 7+7+7+2+2; etc.

A number that is not a prime is called a *composite* number. The first few composite numbers are 4, 6, 8, 9, 10, 12, 14, . . . The prime numbers get more scarce as we consider larger number ranges. For example, there are

168 prime numbers between 1 and 1000

108

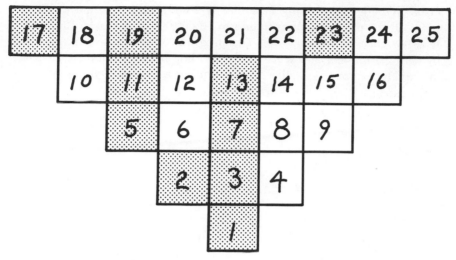

Figure 3-1. The first twenty-five integers.

135 prime numbers between 1000 and 2000
127 prime numbers between 2000 and 3000
120 prime numbers between 3000 and 4000
119 prime numbers between 4000 and 5000

Nevertheless, the list of prime numbers appears to be endless. In fact, it was Euclid who proved that there is an infinite number of primes. Mathematicians have looked in vain for a formula that would generate all the positive prime numbers. Many formulas such as

$$f(n) = n^2 - n + 41$$

will generate prime numbers in the range $1 \leq n \leq 40$ and the formula

$$f(n) = n^2 - 79n + 1601$$

will generate 80 prime numbers when $n = 0, 1, 2, 3, \ldots, 78, 79$. Many other formulas have been developed that will generate prime numbers within a specific range of numbers, but no formula exists that will successively generate primes in any range.

In 1876, the French mathematician Anatole Lucas established a 39-digit prime number.

170,141,183,460,469,231,731,687,303,715,884,105,727

The development of the digital computer provided the mathematician with a tool to help generate much larger prime numbers. A 79-digit prime number was established by the EDSAC in 1952. A 386-digit number was established as a prime in the same year by the SWAC computer.

Prime Number Generating Program. A FORTRAN program that computes all the prime numbers from 3 to 10,000,000 is shown in Figure 3-2. It is flowcharted in Figure 3-3. The program calculates a prime number by taking a number and successively dividing it by all numbers less than the square root of the number. If the number is prime, then the divisor is in the range:

$$3 < d < \sqrt{N}$$

where N is the number.

```
C       PRIME NUMBER GENERATOR

        PRINT 5

00005 FORMAT (36H1PRIME NUMBERS FROM 3 TO 10,000,000 ,///)

        MAYBE = 1

00010 MAYBE = MAYBE + 2

        MAYBER = SQRTF(FLOATF(MAYBE))

        DO 20 IPOSS = 2,MAYBER

        L = MAYBE / IPOSS

        IF ( L * IPOSS - MAYBE) 20,40,50

00020 CONTINUE

        PRINT 30,MAYBE

00030 FORMAT(1H ,10X,I7)

00040 IF (MAYBE - 9999997) 10,10,50

00050 STOP

        END
```

Figure 3-2. Prime number generating program.

If the number cannot be divided evenly by any number less than the square root of the number, then the only divisors are the number 1 and the number itself. Of course this number is a prime number and the program will output this number on the printer. After printing out a prime number, the program attempts to generate another prime. If none is found and the number range exceeds 10,000,000, then the program will stop. The program will output a list of prime numbers similar to that shown in Figure 3-4.

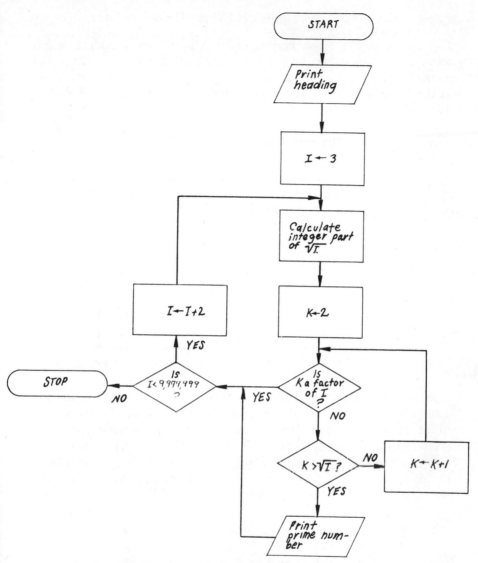

Figure 3-3. Flowchart of the prime number generating program.

3	31	71	109	163	211
5	37	73	113	167	223
7	41	79	127	173	227
11	43	83	131	179	229
13	47	89	137	181	233
17	53	97	139	191	239
19	59	101	149	193	
23	61	103	151	197	●
29	67	107	157	199	●
					●

Figure 3-4. Output of the prime number generating program.

3.2   Sieve of Eratosthenes

Description of the Sieve. Prime numbers have been a subject of great interest since man first became interested in numbers. Of the various methods used for calculating prime numbers, one of the first was known as the *Sieve*, which was attributed to Eratosthenes, a Greek scholar. The *Sieve of Eratosthenes* for the first 100 numbers is shown in Figure 3-5. The scheme consists of writing down the integers in their natural succession and then striking out numbers.

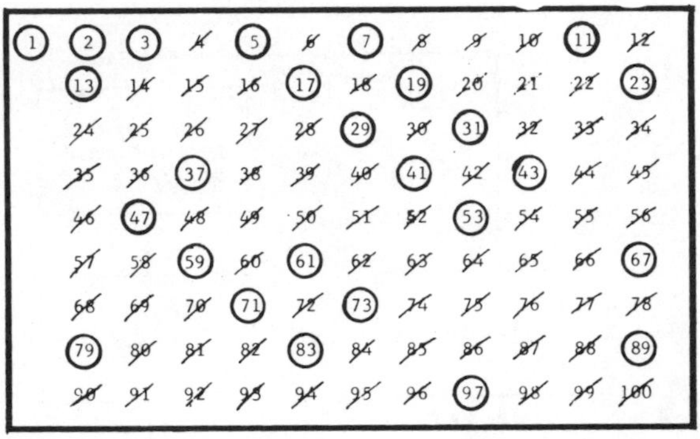

Figure 3-5. The Sieve of Eratosthenes. All Prime Numbers less than 100.

The composite (non-prime) numbers in the number range of the sieve may be eliminated by crossing off, from 2, every second number (4, 6, 8, 10, 12, 14, 16, . . .), then from the next number that is not crossed off, 3, every third number (6, 9, 12, 15, 20, 25, 30, 35, 40, . . .), then from the next number that is not crossed off, 7, every seventh number (14, 21, 28, 35, 42, 49, 56, . . .), and so on. The numbers that were not crossed off are the prime numbers less than 100.

  1 — divisible only by 1
  2 — divisible only by 2 (*the only even prime*)
  3 — not divisible by 2
  5 — not divisible by 2 or 3
  7 — not divisible by 2, 3, or 5
 11 — not divisible by 2, 3, 5, or 7
 13 — not divisible by 2, 3, 5, 7, or 11
 17 — not divisible by 2, 3, 5, 7, 11, or 13
      .
      .
      .
 97 — not divisible by 2, 3, 5, 7, 11, 13, 17, 19, . . ., or 89

The following observations should be noted when using the Sieve of Eratosthenes: (1) Some numbers will be crossed off more than once; (2) It is not necessary to cross off numbers that exceed the square root of the maximum number in the range. In the number range of Figure 3-5 where the maximum number is 100, it was not necessary to cross off the next non-zero number that exceeded 7 since $\sqrt{100} = 10$.

Ingenious as this method of elimination is, it is purely inductive and therefore incapable of proving general properties of prime numbers.

Program. This program first reads a card containing the control value MAX, which specifies the maximum number in Array IPRIME, and then fills the array with the numbers 1, 2, 3, 4, 5, . . ., MAX-2, MAX-1, MAX. After setting all even locations (other than 2) of Array IPRIME equal to zero (IPRIME(4)=0, IPRIME(6)=0, IPRIME(8)=0, IPRIME(10) =0, . . ., etc.) the program computes the square root of the maximum number in the Array

$$\sqrt{MAX}$$

and replaces the variable J with this value. Since we want to calculate all the prime numbers less than MAX, it is not necessary to go beyond the next non-zero number that exceeds $\sqrt{MAX}$. For example, if MAX equalled 1000 it would not be necessary to go beyond the multiples of 31, because $31^2 = 961$ which is the largest square of a prime number less than 1000. The program sets the following elements of Array IPRIME equal to zero.

IPRIME(I) for I = 3, 6, 9, 12, . . .
IPRIME(I) for I = 5, 10, 15, 20, . . .
IPRIME(I) for I = 7, 14, 21, 28, . . .
IPRIME(I) for I = 11, 22, 33, 44, . . .

IPRIME(I) for I = $\sqrt{MAX}$, MAX

After the last non-prime element of Array IPRIME has been set equal to zero, the program prints the contents of the array. The non-zero elements of Array IPRIME are prime numbers. An output printout is shown in Figure 3-8.

A flowchart and FORTRAN program are shown in Figures 3-6 and 3-7.

3.3   Wilson's Prime Number Generator

Description of Wilson's Theorem. Most prime number generating methods involve many division operations. *Wilson's Theorem* is a method that involves only *one division*, but many *multiplications*.

Wilson's Theorem Says:

AN INTEGER (say N) IS A PRIME NUMBER IF AND ONLY IF IT DIVIDES THE NUMBER (N-1) ! + 1

The theorem has little value for use with digital computers due to the large numbers that would be involved, but does provide an interesting computational method.

Program. A flowchart of the Wilson's method of generating prime numbers is shown in Figure 3-9.

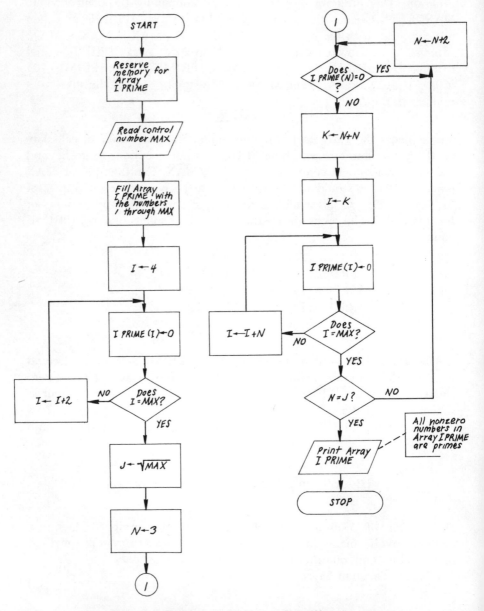

Figure 3-6. Flowchart of the Sieve of Eratosthenes program.

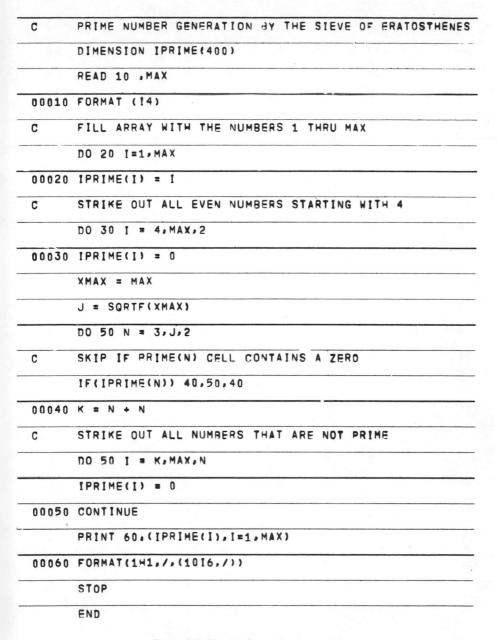

```
C       PRIME NUMBER GENERATION BY THE SIEVE OF ERATOSTHENES

        DIMENSION IPRIME(400)

        READ 10 ,MAX

00010 FORMAT (I4)

C       FILL ARRAY WITH THE NUMBERS 1 THRU MAX

        DO 20 I=1,MAX

00020 IPRIME(I) = I

C       STRIKE OUT ALL EVEN NUMBERS STARTING WITH 4

        DO 30 I = 4,MAX,2

00030 IPRIME(I) = 0

        XMAX = MAX

        J = SQRTF(XMAX)

        DO 50 N = 3,J,2

C       SKIP IF PRIME(N) CELL CONTAINS A ZERO

        IF(IPRIME(N)) 40,50,40

00040 K = N + N

C       STRIKE OUT ALL NUMBERS THAT ARE NOT PRIME

        DO 50 I = K,MAX,N

        IPRIME(I) = 0

00050 CONTINUE

        PRINT 60,(IPRIME(I),I=1,MAX)

00060 FORMAT(1H1,/,(10I6,/))

        STOP

        END
```

Figure 3-7. Sieve of Eratosthenes program.

The program prints out the heading:

PRIME NUMBERS

before reading in the LIMIT card. The LIMIT card specifies the maxi-

1	2	3	0	5	0	7	0	0	0
11	0	13	0	0	0	17	0	19	0
0	0	23	0	0	0	0	0	29	0
31	0	0	0	0	0	37	0	0	0
41	0	43	0	0	0	47	0	0	0
0	0	53	0	0	0	0	0	59	0
61	0	0	0	0	0	67	0	0	0
71	0	73	0	0	0	0	0	79	0
0	0	83	0	0	0	0	0	89	0
0	0	0	0	0	0	97	0	0	0
101	0	103	0	0	0	107	0	109	0
0	0	113	0	0	0	0	0	0	0
0	0	0	0	0	0	127	0	0	0
131	0	0	0	0	0	137	0	139	0
0	0	0	0	0	0	0	0	149	0
151	0	0	0	0	0	157	0	0	0
0	0	163	0	0	0	167	0	0	0
0	0	173	0	0	0	0	0	179	0
181	0	0	0	0	0	0	0	0	0
191	0	193	0	0	0	197	0	199	0
0	0	0	0	0	0	0	0	0	0
211	0	0	0	0	0	0	0	0	0
0	0	223	0	0	0	227	0	229	0
0	0	233	0	0	0	0	0	239	0
241	0	0	0	0	0	0	0	0	0
251	0	0	0	0	0	257	0	0	0
0	0	263	0	0	0	0	0	269	0
271	0	0	0	0	0	277	0	0	0

Figure 3-8. Output of the Sieve of Eratosthenes program.

281	0	283	0	0	0	0	0	0	0
0	0	293	0	0	0	0	0	0	0
0	0	0	0	0	0	307	0	0	0
311	0	313	0	0	0	317	0	0	0
0	0	0	0	0	0	0	0	0	0
331	0	0	0	0	0	337	0	0	0
0	0	0	0	0	0	347	0	349	0
0	0	353	0	0	0	0	0	359	0
0	0	0	0	0	0	367	0	0	0
0	0	373	0	0	0	0	0	379	0
0	0	383	0	0	0	0	0	389	0
0	0	0	0	0	0	397	0	0	0

Figure 3-8. Output of the Sieve of Eratosthenes program (cont'd).

mum possible prime number that will be considered by the program. Care must be taken by the programmer in establishing the limit as the program variables A and PRIME become very large numbers as the value of N increases.

N	PRIME
2	2
3	3
4	7
5	25
6	121
7	721
8	5041
12	39,916,801
16	1,307,674,368,001
19	6,402,373,705,728,001
21	2,432,902,008,176,640,001

The previous illustration of the variables N and PRIME indicate how rapidly the variable PRIME increases as N becomes larger. If the reader cares to examine the values of the variable PRIME more carefully, he will find that PRIME $= (N-1)! + 1$. Once the program determines the value of PRIME, the one division is performed. If the remainder of PRIME/P is zero, then N is a prime number and the program will print

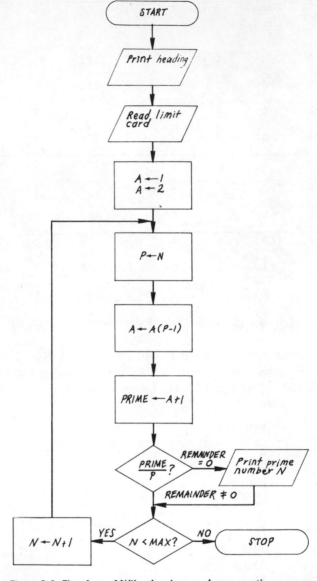

Figure 3-9. Flowchart of Wilson's prime number generating program.

N. If the remainder of the division was not zero, the program goes on to examine the next number to be considered. The program terminates when N equals the Limit Value. Figure 3-10 illustrates program variables for a few values of N.

The FORTRAN program shown in Figure 3-11 uses a value of 11 for MAX and the program output appears as follows

PRIME NUMBERS

2
3
5
7
11

3.4   Prime Number Polynomial

Description of Prime Number Polynomials. Euclid proved that there is an infinite number of primes. Mathematicians have looked, but in vain, for a formula that would generate all the positive prime numbers. The formula $f(n) = n^2 - n + 41$ will generate all the primes in the range $1 \leq n \leq 40$.

$f(n)$	$n^2 - n + 41$
1	$1^2 - 1 + 41 = 41$
2	$2^2 - 2 + 41 = 43$
3	$3^2 - 3 + 41 = 47$
4	$4^2 - 4 + 41 = 53$
5	$5^2 - 5 + 41 = 61$
.	.
.	.
23	$23^2 - 23 + 41 = 547$
.	.
.	.
40	$40^2 - 40 + 41 = 1601$

The formula $f(n) = n^2 - 79n + 1601$ will generate 80 primes when $n = 0, 1, 2, 3, \ldots, 78, 79$. The formula $f(n) = n^2 + n + 17$ will generate

N	P	A	Prime	$\dfrac{\text{Prime}}{P}$	
2	2.	1.	2.	R = 0	Prime
3	3.	2.	3.	R = 0	Prime
4	4.	6.	7.	R ≠ 0	Composite
5	5.	24.	25.	R = 0	Prime
6	6.	120.	121.	R ≠ 0	Composite
7	7.	720.	721.	R = 0	Prime
8	8.	5040.	5041.	R ≠ 0	Composite
9	9.	40320.	40321.	R ≠ 0	Composite
10	10.	362880.	362881.	R ≠ 0	Composite
11	11.	3628800.	3628801.	R = 0	Prime
12	12.	39916800.	39916801.	R ≠ 0	Composite

Figure 3-10. Computed values for Wilson's prime number generating program.

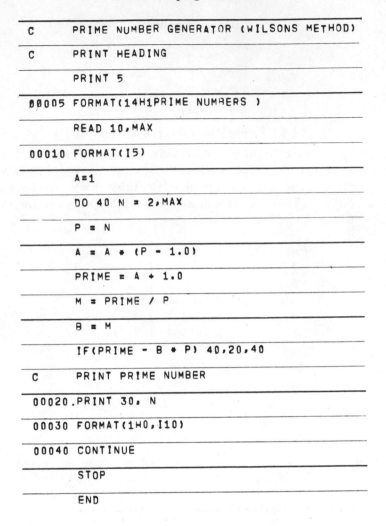

```
C        PRIME NUMBER GENERATOR (WILSONS METHOD)

C        PRINT HEADING

         PRINT 5

00005 FORMAT(14H1PRIME NUMBERS )

         READ 10,MAX

00010 FORMAT(I5)

         A=1

         DO 40 N = 2,MAX

         P = N

         A = A * (P - 1.0)

         PRIME = A + 1.0

         M = PRIME / P

         B = M

         IF(PRIME - B * P) 40,20,40

C        PRINT PRIME NUMBER

00020 .PRINT 30, N

00030 FORMAT(1H0,I10)

00040 CONTINUE

         STOP

         END
```

Figure 3-11. Wilson's prime number generating program.

16 prime numbers when n = 0, 1, 2, 3, ..., 14, 15. The formula $f(n) = n^2+n+41$ will generate 40 prime numbers when n = 0, 1, 2, ...

Program. A FORTRAN program and the program output listing are shown in Figures 3-12 and 3-13, respectively. The program generates 40 prime numbers.

Another Program. The $X^2-79X+1601$ polynomial will generate 80 consecutive prime numbers. The FORTRAN program shown in Figure 3-14 will produce the output shown in Figure 3-15.

```
C       PRIME NUMBER POLYNOMIAL (X**2 -X + 41)

        PRINT 10

00010 FORMAT(58H1THE POLYNOMIAL X**2 -X + 41 WILL PRODUCE 40

 PRIME NUMBE   1RS,//,13X,11HVALUES OF X,10X,13HPRIME NUMBERS,/)

        N=1

00020 NPRIME = N**2 - N + 41

        PRINT 30,N,NPRIME

00030 FORMAT(1H ,15X,I6,16X,I6)

        N=N+1

        IF(N-40)20,20,40

00040 STOP

        END
```

Figure 3-12. FORTRAN program for the X^2-X+41 polynomial.

```
    THE POLYNOMIAL X**2 -X + 41 WILL PRODUCE 40 PRIME NUMBERS

 VALUES OF X   PRIME NUMBERS      VALUES OF X   PRIME NUMBERS
```

VALUES OF X	PRIME NUMBERS	VALUES OF X	PRIME NUMBERS
1	41	21	461
2	43	22	503
3	47	23	547
4	53	24	593
5	61	25	641
6	71	26	691
7	83	27	743
8	97	28	797
9	113	29	853
10	131	30	911
11	151	31	971
12	173	32	1033
13	197	33	1097
14	223	34	1163
15	251	35	1231
16	281	36	1301
17	313	37	1373
18	347	38	1447
19	383	39	1523
20	421	40	1601

Figure 3-13. Output from the X^2-X+41 prime number polynomial program.

```
C       PRIME NUMBER POLYNOMIAL (X**2 - 79X + 1601)

        PRINT 100

00100 FORMAT(74H1THE POLYNOMIAL X**2 -79X + 1601 WILL PRODUCE 80 CONSECU
      1TIVE PRIME NUMBERS,//,13X,11HVALUES OF X,10X,13HPRIME NUMBERS,/)

        K=0

00200 KPRIME = K**2 - 79*K + 1601

        PRINT 300,K,KPRIME

00300 FORMAT(1H ,16X,I4,18X,I4)

        K=K+1

        IF(K-79)200,200,400

00400 STOP

        END
```

Figure 3-14. FORTRAN program for the $X^2 - 79X + 1601$ polynomial.

THE POLYNOMIAL X**2 -79X + 1601 WILL PRODUCE 80 CONSECUTIVE PRIME NUMBERS

VALUES OF X	PRIME NUMBERS	VALUES OF X	PRIME NUMBERS
0	1601	40	41
1	1523	41	43
2	1447	42	47
3	1373	43	53
4	1301	44	61
5	1231	45	71
6	1163	46	83
7	1097	47	97
8	1033	48	113
9	971	49	131
10	911	50	151
11	853	51	173
12	797	52	197
13	743	53	223
14	691	54	251
15	641	55	281
16	593	56	313
17	547	57	347
18	503	58	383
19	461	59	421
20	421	60	461
21	383	61	503
22	347	62	547
23	313	63	593
24	281	64	641
25	251	65	691
26	223	66	743
27	197	67	797
28	173	68	853
29	151	69	911
30	131	70	971
31	113	71	1033
32	97	72	1097
33	83	73	1163
34	71	74	1231
35	61	75	1301
36	53	76	1373
37	47	77	1447
38	43	78	1523
39	41	79	1601

Figure 3-15. Output of the $X^2 - 79X + 1601$ prime number polynomial program.

CHAPTER 4

Miscellaneous Game Programs

4.1 Pick-A-Number Game

Game Description. THIS GAME is an application of the *binary number system.* A set of cards is prepared using binary notation. A set of cards for a *4 card Pick-A-Number* game is shown in Figure 4-1.

To play this game with the above set of cards, one asks someone to think of a number (between 1 and 15) and to indicate the card or cards on which the number appears. If the number appeared on card A *only,* the number would be 1. If the number appeared on *card A, card B* and *card C,* then the sum of the numbers in the *top left corner* of each card would determine that the correct number is 7. Ten would be the number if the number were contained on *cards B* and *D* because $8+2 = 10$. It is easily seen that one needs only to add the numbers in the *top left squares* of each card to predict the chosen number.

Of course the secret of this trick is in the making of the cards. Think of each card as representing a power of 2 ($2^0, 2^1, 2^2, 2^4$). In other words, each card will represent a decimal value equal to some power of 2.

CARD A represents the value 2^0 or 1.
CARD B represents the value 2^1 or 2.
CARD C represents the value 2^2 or 4.
CARD D represents the value 2^3 or 8.

Card D		Card C		Card B		Card A	
8	12	4	12	2	10	1	9
9	13	5	13	3	11	3	11
10	14	6	14	6	14	5	13
11	15	7	15	7	15	7	15

Figure 4-1. A 4 card Pick-A-Number card set.

123

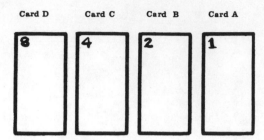

Figure 4-2. Card values for the Pick-A-Number program.

Place the above values on the cards as shown in Figure 4-2.

Convert the decimal numbers 1 through 15 to their binary representation. Figure 4-3 may be referenced if the reader is not familiar with the binary number system.

Decimal Number	Binary Number			
	2^3	2^2	2^1	2^0
1	0	0	0	1
2	0	0	1	0
3	0	0	1	1
4	0	1	0	0
5	0	1	0	1
6	0	1	1	0
7	0	1	1	1
8	1	0	0	0
9	1	0	0	1
10	1	0	1	0
11	1	0	1	1
12	1	1	0	0
13	1	1	0	1
14	1	1	1	0
15	1	1	1	1

Figure 4-3. Decimal/Binary conversion chart.

If a 1 appears in the converted number, then it should also appear on the respective card.

Let us use the decimal value 9 as an example. The decimal value 9 should appear on cards A and D since the binary value for a decimal 9 contains two ones: one in the position that specifies Card D and one in the position that specifies Card A. The other decimal values are represented on the cards in a similar manner.

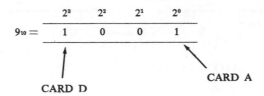

As seen in the previous example, the four card game is limited to the number range 1 through 15. If the card set were expanded to five cards, the number range would be expanded to include the numbers 1 through 31 as shown in Figure 4-4. Using the 5 card set, if the Cards E, C and A were selected, one need only compute 16 +4 +1 to determine that the number is 21.

Program. The Pick-A-Number program will read an input card containing one through five card choices, analyze the card values and print a message of the following form:

> YOUR CHOICE OF CARDS WAS *a, b, c, d, e*
> THE NUMBER IS *k*

where *a, b, c, d* and *e* are the card numbers that were chosen by the player (1, 2, 3, 4, or 5), and *k* is the number calculated by the program.

The program checks for an illegal card number (card number > 5), and if one is found, the following message will be printed.

> CARD *n* is ILLEGAL

where *n* is the card number.

Card E	Card D	Card C
16 20 24 28	8 12 24 28	4 12 20 28
17 21 25 29	9 13 25 29	5 13 21 29
18 22 26 30	10 14 26 30	6 14 22 30
19 23 27 31	11 15 27 31	7 15 23 31

Card B	Card A
2 10 18 26	1 9 17 25
3 11 19 27	3 11 19 27
6 14 22 30	5 13 21 29
7 15 23 31	7 15 23 31

Figure 4-4. A 5 card Pick-A-Number card set.

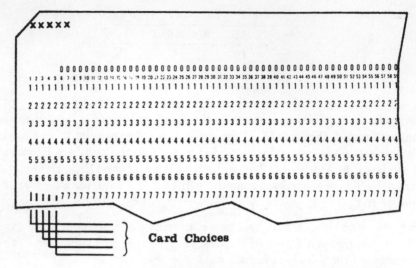

Card Choices

**Card number may appear in any order
(1, 2, 4, 0, 0 is identical to 4, 1, 0, 2, 0 or 0, 0
2, 4, 1)**

Figure 4-5. Input card for the Pick-A-Number program.

The format of the input number card is shown in Figure 4-5.

Figure 4-6 illustrates a flowchart of the Pick-A-Number program and a FORTRAN program is shown below.

```
C       PICK--A--NUMBER   PROGRAM
C       INITIALIZE COUNTER
        DIMENSION KARD (5)
        KTR = 0
C       READ CARD CHOICES
        READ 20,(KARD(I),I=1,5)
00020 FORMAT (5I1)
        PRINT 30,(KARD(I),I=1,5)
00030 FORMAT(26H1 YOUR CHOICE OF CARDS WAS ,5I5 )
        DO 60 K=1,5
        DO 40 J=1,5
C       JUMP TO 60 IF CARD VALUE IS ZERO
        IF (KARD(K))35,60,35
C       JUMP TO 36 IF CARD VALUE IS GREATER THAN 5
00035 IF (KARD(K) - 5)38,36,36
00036 PRINT 37, K
```

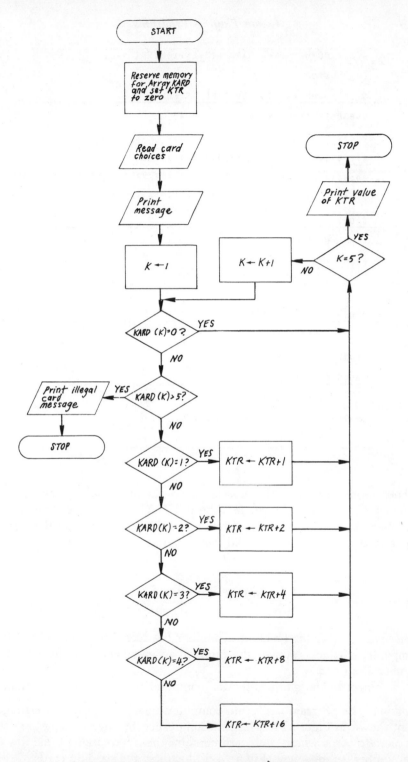

Figure 4-6. Flowchart of the Pick-A-Number program.

```
00037 FORMAT (6H0 CARD ,I3,2X,10HIS INVALID )
      STOP
C     JUMP TO 50 WHEN CARD IS FOUND
00038 IF (KARD(K) - J) 40,50,40
00040 CONTINUE
00050 GO TO (51,52,53,54,55),J
00051 KTR = KTR + 1
      GO TO 60
00052 KTR = KTR + 2
      GO TO 60
00053 KTR = KTR + 4
      GO TO 60
00054 KTR = KTR + 8
      GO TO 60
00055 KTR = KTR + 16
00060 CONTINUE
      PRINT 70, KTR
00070 FORMAT(16H0 THE NUMBER IS  ,I2)
      STOP
      END
```

Four input cards and the associated output produced by the Pick-A-Number program are shown in Figure 4-7. The first three input cards contained legal card numbers (numbers 1 through 5) while the fourth input card contained the invalid card number 9.

4.2 Binary Number Game

Description of Game. Think of a number between 1 and 63 and tell the computer in which rows of the number arrangement of Figure 4-8 the number lies.

The object of the game is for the computer to determine the number.

Program. The program reads into computer memory the number arrangement shown in Figure 4-8. This is accomplished by keypunching the data on six data cards where each card contains the information of one row of the number arrangement. Figure 4-9 illustrates the six data cards.

INPUT CARD VALUE	PICK-A-NUMBER PROGRAM OUTPUT
12345	YOUR CHOICE OF CARDS WAS 1,2,3,4,5 THE NUMBER IS 31
34000	YOUR CHOICE OF CARDS WAS 3,4 THE NUMBER IS 12
25001	YOUR CHOICE OF CARDS WAS 2,5,1 THE NUMBER IS 19
12903	CARD 3 IS ILLEGAL

Figure 4-7. Sample input and output of the Pick-A-Number program.

```
1  32 33 34 35 36 37 38 39 40 41 42 43 44 45 46 47 48 49 50 51 52 53 54 55 56 57 58 59 60 61 62
2   3  6  7 10 11 14 15 18 19 22 23 26 27 30 31 34 35 38 39 42 43 46 47 50 51 54 55 58 59 62 63
4   5  6  7 12 13 14 15 20 21 22 23 28 29 30 31 36 37 38 39 44 45 46 47 52 53 54 55 60 61 62 63
8   9 10 11 12 13 14 15 24 25 26 27 28 29 30 31 40 41 42 43 44 45 46 47 56 57 58 59 60 61 62 63
16 17 18 19 20 21 22 23 24 25 26 27 28 29 30 31 48 49 50 51 52 53 54 55 56 57 58 59 60 61 62 63
32 33 34 35 36 37 38 39 40 41 42 43 44 45 46 47 48 49 50 51 52 53 54 55 56 57 58 59 60 61 62 63
```

Figure 4-8. Number table for the binary number game.

After reading in the data cards the program prints the message

THINK OF A NUMBER BETWEEN 1 AND 63. TELL
ME WHICH ROWS OF THE FOLLOWING TABLE YOU
CAN FIND THE NUMBER IN AND I WILL PRINT
THE NUMBER.

and then prints the number arrangement of Figure 4-8. The computer then *pauses* until the game player inserts a control card in the card reader and again informs the computer to continue. The control card simply specifies to the program what rows the selected number lies in.

A flowchart of the program is shown in Figure 4-10. This program checks to see if the number is in each row and increments the variable counter NUMBER by an appropriate value for each row it finds the number in. If the number is in row 1, NUMBER is incremented by 1; if in row 2, it is incremented by 2; if in row 3, incremented by 4; if in row 4, incremented by 8; if in row 5, incremented by 16; and if in row 6, it is incremented by the value 32. The reader should notice that the

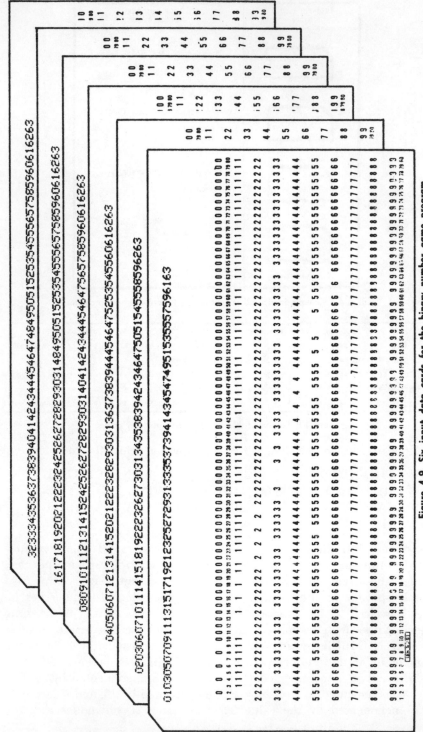

Figure 4-9. Six input data cards for the binary number game program.

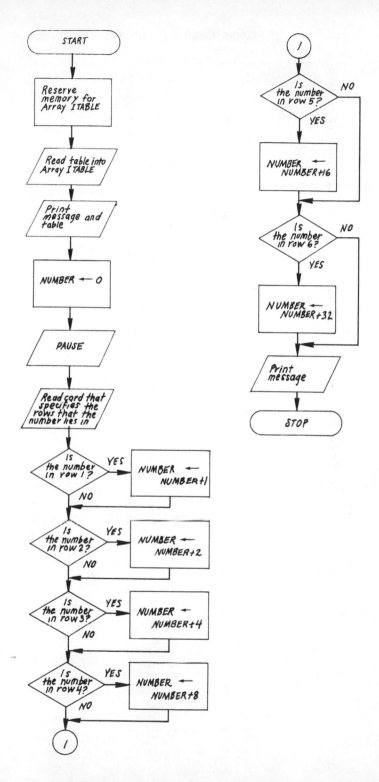

Figure 4-10. Flowchart of the binary number game program.

```
THINK OF A NUMBER BETWEEN 1 AND 63. TELL ME WHICH ROWS OF THE FOLLOWING TABLE YOU CAN FIND THE NUMBER
IN AND I WILL PRINT THE NUMBER.

 1 32 33 34 35 36 37 38 39 40 41 42 43 44 45 46 47 48 49 50 51 52 53 54 55 56 57 58 59 60 61 62

 2  3  6  7 10 11 14 15 18 19 22 23 26 27 30 31 34 35 38 39 42 43 46 47 50 51 54 55 58 59 62 63

 4  5  6  7 12 13 14 15 20 21 22 23 28 29 30 31 36 37 38 39 44 45 46 47 52 53 54 55 60 61 62 63

 8  9 10 11 12 13 14 15 24 25 26 27 28 29 30 31 40 41 42 43 44 45 46 47 56 57 58 59 60 61 62 63

16 17 18 19 20 21 22 23 24 25 26 27 28 29 30 31 48 49 50 51 52 53 54 55 56 57 58 59 60 61 62 63

32 33 34 35 36 37 38 39 40 41 42 43 44 45 46 47 48 49 50 51 52 53 54 55 56 57 58 59 60 61 62 63


YOU PICKED ROWS  1, 3, 5, 6
THE NUMBER IS    53
```

Figure 4-11. Output of the binary number game program.

values 1, 2, 4, 8, 16 and 32 are all in column 1 of the number arrangement of Figure 4-8.

After the program has checked the last row, it causes the following message to be printed.

YOU PICKED ROWS   X X X X X
THE NUMBER IS   N

where the X values indicate the rows that were picked by the game player and N is the number that the computer determined.

The FORTRAN program is shown below.

Figure 4-11 illustrates the output of the program whenever a game player selects rows 1, 3, 5 and 6. The number is 53.

```
C       BINARY NUMBER GAME
        DIMENSION ITABLE(5,32)
C       READ TABLE VALUES INTO ARRAY ITABLE
        READ 10,((ITABLE(K,L),L=1,32),K=1,6)
00010 FORMAT (32I2)
C       PRINT HEADING AND TABLE
        PRINT 20
00020 FORMAT(1H2H1THINK OF A NUMBER BETWEEN 1 AND 63, TELL ME WHICH ROWS
       1 OF THE FOLLOWING TABLE YOU CAN FIND THE NUMBER ,/,32H IN AND I W
       1ILL PRINT THE NUMBER, ,/// )
        PRINT 30,((ITABLE(J,K),K=1,32),J=1,6)
00030 FORMAT(1H0,/,6(32I3,//))
        NUMBER = 0
C       STOP COMPUTER UNTIL OPERATOR READS IN INPUT CARD
C       SPECIFYING THE ROWS IN WHICH THE NUMBER LIES IN.
        PAUSE
C       READ INPUT CARD THAT SPECIFIES THE ROWS THAT
C       THE NUMBER LIES IN.
        READ 40,ROW1,ROW2,ROW3,ROW4,ROW5,ROW6
00040 FORMAT(6F5.0)
C       DOES THE NUMBER LIE IN ROW 1
        IF(ROW1)180,60,50
00050 NUMBER = NUMBER + 1
C       DOES THE NUMBER LIE IN ROW 2
00060 IF(ROW2)180,80,70
00070 NUMBER = NUMBER + 2
C       DOES THE NUMBER LIE IN ROW 3
00080 IF(ROW3)180,100,90
00090 NUMBER = NUMBER + 4
C       DOES THE NUMBER LIE IN ROW 4
```

```
00100 IF(ROW4)100,120,110
00110 NUMBER = NUMBER + 8
C      DOES THE NUMBER LIE IN ROW 5
00120 IF(ROW5)100,140,130
00130 NUMBER = NUMBER + 16
C      DOES THE NUMBER LIE IN ROW 6
00140 IF(ROW6)100,160,150
00150 NUMBER = NUMBER + 32
C      PRINT THE NUMBER
00160 PRINT 170,ROW1,ROW2,ROW3,ROW4,ROW5,ROW6,NUMBER
00170 FORMAT(1H0,/,16H YOU PICKED ROWS ,6F4.0,/,14H THE NUMBER IS ,I7)
00180 STOP
      END
```

4.3   Counterfeit Coin Game

Description of Game. Nine coins are laying on a table. One of them is counterfeit and is lighter in weight than the other eight. What is the least number of weighings (using a pair of scales) that is required to determine which coin is counterfeit?

This problem may be solved by dividing the nine coins into three separate piles, say PILE1, PILE2, and PILE3. Place any two of the piles of coins on the scale and determine which, if any, pile is lighter. For example, the following conditions would exist if PILE1 and PILE2 were placed on the scale, as seen in Figure 4-12.

- Counterfeit coin is in PILE1 if PILE1 is lighter than PILE2 (PILE1 < PILE2).
- Counterfeit coin is in PILE2 if PILE1 is heavier than PILE2 (PILE1 > PILE2).
- Counterfeit coin is in PILE3 if PILE1 and PILE2 weigh the same (PILE1 = PILE2).

After the pile containing the counterfeit coin is determined, it is necessary to find out which of the three coins is the counterfeit coin. This is accomplished by weighing the coins in a similar manner. For example, if the coins were labeled COIN1, COIN2 and COIN3, and COIN1 and COIN2 were placed on the scale, the following relationships would determine which of the three coins is counterfeit.

COIN1 is counterfeit if COIN1 < COIN2
COIN2 is counterfeit if COIN1 > COIN2
COIN3 is counterfeit if COIN 1 = COIN2

Extension of Game. In cases where the number of coins is different from nine, it is sufficient to divide the coins into k piles with 3^{k-1} coins

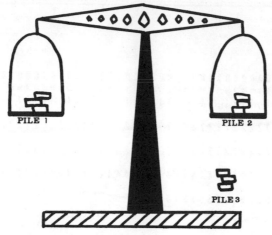

Figure 4-12. Scale weighing of coins in the counterfeit coin game.

in each pile. Of course this division assumes that the original pile contained 3^k coins.

If the original pile contained a number of coins less than 3^k but greater than 3^{k-1}, then the counterfeit lighter coin may also be identified in k balancings. For example, if we were asked to find a counterfeit coin in a pile of 12 coins, the coins could be divided into three piles:

> PILE1 contains coins 1 2 3 4
> PILE2 contains coins 5 6 7 8
> PILE3 contains coins 9 10 11 12

If PILE1 is lighter than PILE2, then the lighter coin is in PILE1. If PILE1 > PILE2, then the lighter coin is in PILE2. If PILE1 = PILE2, then the counterfeit coin is contained in PILE3. Once determining the pile that contains the counterfeit coin, it is a simple task to identify the coin that differs in weight from the other three coins.

Program. Nine coins are represented in Array KOIN. If an element of the array has a value of 1, it is assumed to represent a good coin. If the element is zero, the coin is assumed to be counterfeit. The program reads into memory a punched card containing the coin values, as seen in Figure 4-13, and determines which one of the nine coins is the counterfeit coin.

Once the program determines which coin is counterfeit, the message

COUNTERFEIT COIN IS COIN n

is printed on the high-speed printer. The value of n will depend upon which coin is found to be counterfeit. A flowchart is shown in Figure 4-14. The FORTRAN program is as follows.

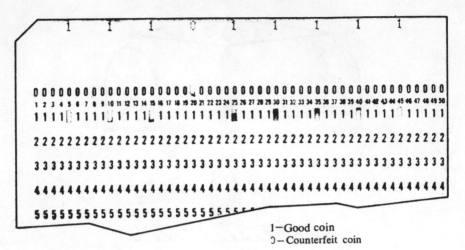

Figure 4-13. Input card for the counterfeit coin program. Counterfeit coin is the fourth coin.

```
C        COUNTERFEIT  COIN  PROBLEM
C        RESERVE MEMORY FOR ARRAY COIN
         DIMENSION KOIN(9)
C        READ COIN VALUES INTO MEMORY
         READ 10,(KOIN(K),K=1,9)
      10 FORMAT(9I5)
C        GROUP COINS INTO THREE PILES
         IPILE1 = KOIN(1) + KOIN(2) + KOIN(3)
         IPILE2 = KOIN(4) + KOIN(5) + KOIN(6)
         IPILE3 = KOIN(7) + KOIN(8) + KOIN(9)
C        DETERMINE WHICH PILE CONTAINS
C        THE COUNTERFEIT COIN
          IF(IPILE1 -IPILE2) 20,100,60
C '      PILE1 CONTAINS THE COUNTERFEIT COIN
      20 IF(KOIN(1) - KOIN(2)) 30,50,40
C        COIN1 IS THE COUNTERFEIT COIN
      30 I = 1
         GO TO 140
C        COIN2 IS THE COUNTERFEIT COIN
      40 I = 2
         GO TO 140
C        COIN3 IS THE COUNTERFEIT COIN
      50 I = 3
         GO TO 140
C        PILE2 CONTAINS THE COUNTERFEIT COIN
      60 IF(KOIN(4) - KOIN(5)) 70,90,80
C        COIN4 IS THE COUNTERFEIT COIN
      70 I = 4
         GO TO 140
C        COIN5 IS THE COUNTERFEIT COIN
      80 I = 5
         GO TO 140
```

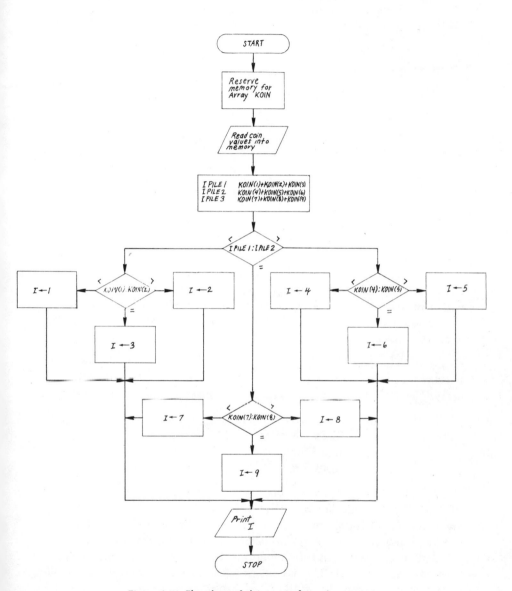

Figure 4-14. Flowchart of the counterfeit coin program.

```
C        COIN6 IS THE COUNTERFEIT COIN
    90   I = 6
         GO TO 140
C        PILE3 CONTAINS THE COUNTERFEIT COIN
   100   IF(KOIN(7) - KOIN(9)) 110,130,120
C        COIN7 IS THE COUNTERFEIT COIN
   110   I = 7
         GO TO 140
C        COIN8 IS THE COUNTERFEIT COIN
   120   I = 8
         GO TO 140
C        COIN9 IS THE COUNTERFEIT COIN
   130   I = 9
   140   PRINT 150,I
   150   FORMAT(25H COUNTERFEIT COIN IS COIN,I3)
         STOP
         END
```

4.4   Checkers and Kings

Description of Game. A standard checkerboard has 64 squares upon which *red checkers, black checkers, red kings* or *black kings* may reside. The object of the game is to determine how many of each checker type is on the checkerboard.

Program. This program uses an array called KKK to represent the checkerboard and the subscripts I and J to locate a position within the array. Each of the array elements could represent one of the following:

$$
\begin{array}{llll}
\text{Red Checker} & - & 10 & - & \text{IRC} \\
\text{Black Checker} & - & 20 & - & \text{IBC} \\
\text{Red King} & - & 30 & - & \text{IRK} \\
\text{Black King} & - & 40 & - & \text{IBK}
\end{array}
$$

The value following the checker type is used by the program to identify the type of checker, and the variable names following the values are used by the program to store the count of each checker type. The program generates a count for each checker type by examining each element of the array and incrementing the respective counts each time a checker is detected. The flowchart of Figure 4-15 describes the flow of the program and the FORTRAN program is shown in Figure 4-16.

Figure 4-17 illustrates a printout of the program. Following the printout of the 8 by 8 Array are the values of the counters IRC, IBC, IRK and IBK thus indicating that there were 4 red checkers, 6 black checkers, 3 red kings and 2 black kings specified in Array KKK.

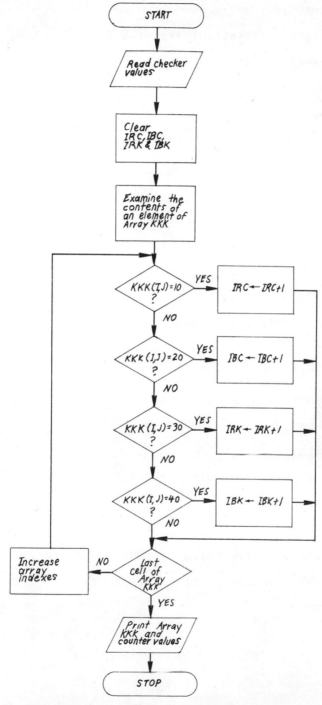

Figure 4-15. Flowchart of the checkers and kings program.

```
         DIMENSION  KKK(8,8)

         READ 100,((KKK(I,J),I=1,8),J=1,8)

00100 FORMAT (32I2)

         IRC = 0

         IBC = 0

         IRK = 0

         IBK = 0

         DO 900 I=1,8

         DO 900 J=1,8

         IF (KKK(I,J) - 10)300,200,300

00200 IRC = IRC + 1

         GO TO 900

00300 IF (KKK(I,J) - 20)500,400,500

00400 IBC = IBC + 1

         GO TO 900

00500 IF (KKK(I,J) - 30)700,600,700

00600 IRK = IRK + 1

         GO TO 900

00700 IF (KKK(I,J) - 40)900,800,900

00800 IBK = IBK + 1

00900 CONTINUE

         PRINT 1000,((KKK(I,J),J=1,8),I=1,8),IRC,IBC,IRK,IBK

01000 FORMAT (1H1,////,8(8I5,//),////,4I10)

         STOP

         END
```

Figure 4-16. Checkers and kings program.

0	0	0	0	40	0	0	0
10	30	0	0	0	40	0	0
0	10	0	0	20	30	0	0
0	0	0	0	0	0	0	0
0	0	0	0	10	20	0	0
10	0	0	0	0	0	0	0
20	0	0	0	20	0	0	0
0	20	0	0	20	30	0	0

4	6	3	2

Figure 4-17. Output of the checkers and kings program.

CHAPTER 5

Puzzles

5.1 The 15 Puzzle

Description of Puzzle. The 15 Puzzle is a 4 by 4 square array consisting of the numbers 1 through 15 and one blank position as shown in Figure 5-1.

The object of the game is to rearrange the number configuration of Figure 5-1 to some other configuration (for example, Figure 5-2).

The 15 numbered squares are free to slide about within a box which is usually made of plastic, wood or metal. The problem is to slide the squares without removing them from the box until the new number configuration is achieved. This is accomplished by sliding a numbered square into the blank position. It is readily seen in Figure 5-1 that the only possible first moves would be the 12 or 15.

The 15 Puzzle, which is also referred to as the *Boss Puzzle* or the *Jeu de Taquin,* was one of the many puzzles invented by Sam Loyd. For several years after its appearance in 1878, the puzzle was extremely popular in Europe and even today the puzzle can be bought in almost any American department, dime or drug store. In the 1800's, the game

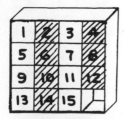

Figure 5-1. Normal position of the 15 puzzle (horizontal position).

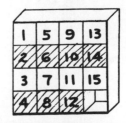

Figure 5-2. Vertical position of the 15 puzzle.

142

was played on all levels of society—in the factories, salons, palaces, streets and restaurants. Today the puzzle is played by many classes of people and modified versions of the puzzle have been favorites of children for many years.

Shortly after the puzzle was introduced, large cash prizes were offered for solutions to apparently simple arrangements. Many people responded to these offers with solutions, but no one could recall his moves well enough to collect a prize. If any one had collected a prize, he would have certainly been cheating, because it was later learned that various number arrangements were impossible to achieve. Two American mathematicians proved that approximately *10 trillion arrangements were impossible to achieve.* We can now see how generous prizes could be offered by Sam Loyd and others for the impossible arrangements of the 15 Puzzle. Would-be cheaters are foiled today when playing this puzzle as the number squares can be moved vertically or horizontally within the box, but they cannot be removed from the box. Many impossible solutions would have been reduced if this contraption had existed in Germany and France during the 19th century.

There are 16 squares in a 4 by 4 square and only 15 numbered squares to move about in this square box. This establishes the table of arrangements shown in Table 5-1.

Table 5-1. Possible and Impossible Arrangements of the 15 Puzzle.

Different arrangements	20,922,789,888,000 (16!)
Possible arrangements	10,461,394,944,000
Impossible arrangements	10,461,394,944,000

As seen in Table 5-1, there are as many impossible number arrangements as there are possible ones (approximately 10 trillion).

Rather than get. involved with the derivation of the method of determining whether a specific number arrangement is possible or not, I will simply state the method. In the normal position, Figure 5-1, every number block appears in its proper numerical order (i.e., no number precedes a number smaller than itself). The order of the numbers will change when a new number arrangement exists. By recording how many times a number precedes one smaller than itself, and whether this value is odd or even, will determine if the new number configuration is possible or impossible to achieve. Table 5-2 describes this method in detail.

Table 5-2. Rules for Determining if a Specific 15 Puzzle Arrangement is Possible or Impossible to Obtain.

1. Let N be a number in position A (Reference Figure 5-3) of the array to be achieved. Count how many numbers smaller than N are in positions higher lettered than A. Count the blank as 16.
2. Do this for all 16 positions (A through P), and add up the count.
3. If the blank square is one of the shaded squares of Figure 5-3 (B, D, E, G, J, L, M, or O), add 1 to the sum. Do not change the sum if the blank square was one of the unshaded squares (A, C, F, H, I, K, N, or P).
4. The new array is POSSIBLE if the sum is EVEN.
5. The new array is IMPOSSIBLE if the sum is ODD.

Figure 5-3. Lettered position of working square.

Proof that the array in Figure 5-4 can be achieved and that the arrangement in Figure 5-5 is impossible is used to illustrate the method described in Table 5-2.

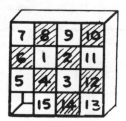

Figure 5-4. Possible arrangement of the 15 puzzle.

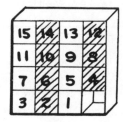

Figure 5-5. Impossible arrangement of the 15 puzzle.

Table 5-3 illustrates the sum of the counts for positions A through P. Since the total sum of Figure 5-4 is even (42), the array is possible to achieve. The array in Figure 5-5 is one of the impossible arrangements of the 15 Puzzle (total sum is odd, 105).

Figure 5-6 illustrates another possible-impossible situation of the 15 Puzzle. The arrangement in Figure 5-6a is *possible* since the count = 2 while it may readily be seen that the arrangement in Figure 5-6b is *impossible* because one additional move can change the count from even to odd.

Several other arrangements of the 15 Puzzle are shown in Figure 5-7. The reader should verify the correctness of whether they are possible or impossible arrangements.

Table 5-3. Determining the Sums for the Configurations Shown in Figures 5-4 and 5-5.

Lettered Position	Numbers Smaller than N	
	Figure 5-4 Arrangement	Figure 5-5 Arrangement
A	6	14
B	6	13
C	6	12
D	6	11
E	5	10
F	0	9
G	0	8
H	3	7
I	2	6
J	1	5
K	0	4
L	0	3
M	3	2
N	2	1
O	1	0
P	0	0
Sum of above	41	105
Blank position	+1	+0
Total Sum	42	105

Program. A flowchart of a FORTRAN program that will determine if a given 15 Puzzle arrangement can be achieved is shown in Figure 5-9. Input to the program is a punched card containing the arrangement to be achieved and it must be in the format specified in Figure 5-8.

The program will read in the input card and store this number arrangement in Array NUM. A counter (KO) is established and is used to indicate when the arrangement is possible (EVEN count) or impossible (ODD count). The number arrangement and one of the following messages will be printed on the printer.

• THE ABOVE 15 PUZZLE ARRANGEMENT CAN BE ACHIEVED
• THE ABOVE 15 PUZZLE ARRANGEMENT CANNOT BE ACHIEVED

Of course, the choice of the proper message is dependent upon whether the count is EVEN or ODD.

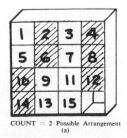

COUNT = 2 Possible Arrangement
(a)

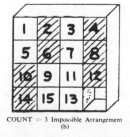

COUNT = 3 Impossible Arrangement
(b)

Figure 5-6. Possible and impossible arrangements of the 15 puzzle.

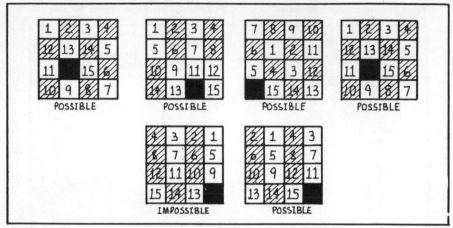

Figure 5-7. Various arrangements of the 15 puzzle.

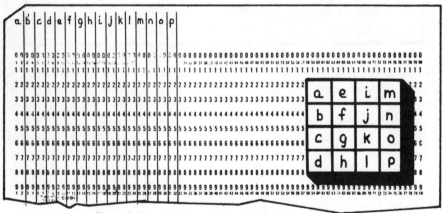

Figure 5-8. Input data card for the 15 puzzle program.

The 15 Puzzle FORTRAN Program is shown in Figure 5-10.

Sample Run Using The 15 Puzzle Program. It is desired to determine if the number arrangement of Figure 5-11 is possible to achieve.

An input card is prepared and it will look like the one shown in Figure 5-12. Note that a 16 was used to indicate the blank square.

The program will read the punched card and use the information to establish a count. The count is contained in the variable KO. Table 5-4 illustrates how the count progresses as the program determines if the number arrangement is possible or impossible.

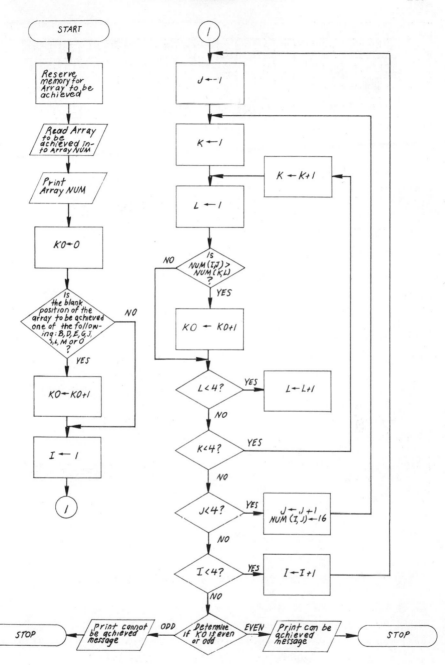

Figure 5–9. Flowchart of the 15 puzzle program.

```
C       15 PUZZLE
        DIMENSION NUM(4,4)
        READ 1,((NUM(I,J),I=1,4),J=1,4)
00001 FORMAT (16I2)
        PRINT 14, ((NUM(I,J),J=1,4),I=1,4)
00014 FORMAT (1H1,/,4(10X,4I5,///),///)
        KO = 0
        IF (NUM(1,2) - 15) 2,9,2
00002 IF (NUM(1,4) - 16) 3,9,3
00003 IF (NUM(2,1) - 16) 4,9,4
00004 IF (NUM(2,3) - 16) 5,9,5
00005 IF (NUM(3,2) - 16) 6,9,6
00006 IF (NUM(3,4) - 16) 7,9,7
00007 IF (NUM(4,1) - 16) 8,9,8
00008 IF (NUM(4,3) - 16)10,9,10
00009 KO = KO + 1
00010 DO 13 I = 1,4
        DO 13 J = 1,4
        DO 12 K = 1,4
        DO 12 L = 1,4
        IF (NUM(I,J) - NUM(K,L)) 12,12,11
00011 KO = KO + 1
00012 CONTINUE
00013 NUM(I,J) = 16
        N = KO / 2
        IF (KO - 2*N) 17,15,17
00015 PRINT 16
00016 FORMAT (50H0 THE ABOVE 15 PUZZLE ARRANGEMENT CAN BE ACHIEVED     )
        GO TO 19
00017 PRINT 18
00018 FORMAT (53H0 THE ABOVE 15 PUZZLE ARRANGEMENT CANNOT BE ACHIEVED )
00019 STOP
        END
```

Figure 5-10. FORTRAN program for the 15 puzzle.

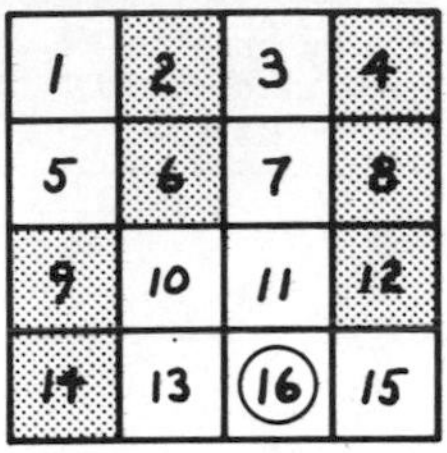

Figure 5-11. Arrangement to be achieved.

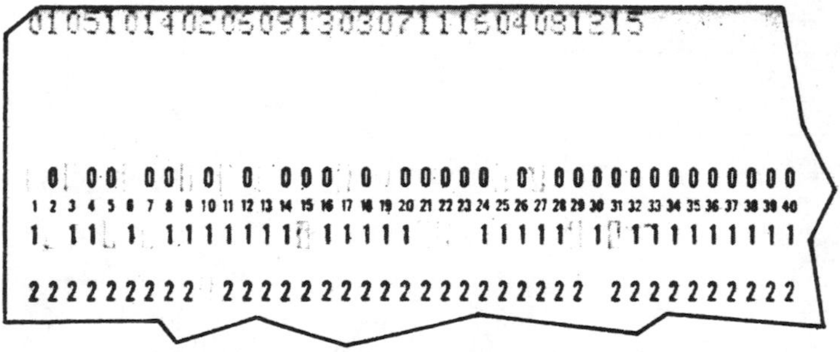

Figure 5-12. Input card for the array of Figure 5-11.

Table 5-4. Determining the Count of the Arrangement Shown in Figure 5-11.

KO	I	J	K	L	Comments	
1	1				Blank square is 0	
2	3	1	3	2	NUM(3,1)>NUM(3,2)	10>9
3	4	1	4	2	NUM(4,1)>NUM(4,2)	14>13
4	4	3	4	4	NUM(4,3)>NUM(4,4)	16>15

The arrangement of Figure 5-11
is possible because KO is EVEN

Output of the 15 Puzzle program is shown in Figure 5-13.

1	2	3	4
5	6	7	8
10	9	11	12
14	13	16	15

THE ABOVE 15 PUZZLE
ARRANGEMENT CAN BE
ACHIEVED

Figure 5-13. Output of the 15 puzzle program.

Other Puzzle Arrangements. Figure 5-14 illustrates several 15 puzzle
number arrangements and the associated value of the program variable

15 Puzzle Arrangement	Value of KO	Configuration *CAN* or *CANNOT* be achieved
1 2 3 4 / 5 6 7 8 / 9 10 11 12 / 13 14 15 16	0	CAN
1 2 3 4 / 5 6 7 8 / 10 9 11 12 / 14 13 16 15	4	CAN
1 5 9 13 / 2 6 10 14 / 3 7 11 15 / 4 8 12 16	36	CAN
4 3 2 1 / 8 7 6 5 / 12 11 10 9 / 15 14 13 16	21	CANNOT
7 8 9 10 / 6 1 2 11 / 5 4 3 12 / 16 15 14 13	42	CAN
1 2 3 4 / 5 6 7 8 / 10 9 11 12 / 14 15 13 16	3	CANNOT
15 14 13 12 / 11 10 9 8 / 7 6 5 4 / 3 2 1 16	105	CANNOT
1 2 3 4 / 5 6 7 8 / 10 9 11 12 / 14 13 15 16	2	CAN

Figure 5-14. 15 puzzle arrangements and associated values of KO.

KO. The count represented by this variable is an indication of whether the number configuration *can* or *cannot* be achieved.

5.2 The 34 Puzzle

Discussion of Puzzle. The 34 Puzzle is an arrangement of 16 numbers in a 4 by 4 square in such a way that the sum of each horizontal row, vertical column, main diagonal and group of four adjacent numbers is equal to 34. The puzzle is played using a standard 15 *puzzle box* and the numbers are moved within the box to form the 34 puzzle number arrangement. The blank space in the puzzle box is assigned a value of 16.

The 34 Puzzle arrangement of Figure 5-15 is used to illustrate how to determine if a number arrangement meets the requirements of the puzzle.

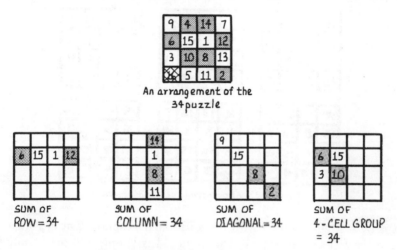

An arrangement of the
34 puzzle

Figure 5-15. 34 puzzle number arrangement.

Figure 5-16 illustrates a number arrangement that is not a valid arrangement of the 34 Puzzle. The invalid number arrangement of Figure 5-16 was arrived at by interchanging the values 6 and 9 of the valid arrangement of Figure 5-15. Figure 5-16 also shows where the sum of row 1 and 2, diagonal 1 and one 4-cell group sum does not equal 34.

Several valid number arrangements of the 34 Puzzle are shown in Figure 5-17. Other arrangements are possible and it is suggested that the reader determine a few of these arrangements in order to have a better understanding of the puzzle.

If the reader is familiar with the properties of *Magic Squares* (see Chapter 2), then it should be evident that a solution to the 34 Puzzle

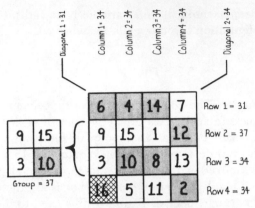

Figure 5-16. Invalid arrangement of the 34 puzzle.

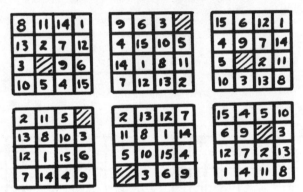

Figure 5-17. Arrangements of the 34 puzzle.

is a 4 by 4 Magic Square with the additional property that any adjacent four cells will also equal 34. The value 34 is simply the magic number of the square:

$$\text{Magic Number} = \frac{N(N^2+1)}{2} = \frac{4(16+1)}{2} = \frac{68}{2} = 34$$

where N is the number of cells in one side of the square.

Program. The 34 Puzzle Checking Program reads into computer storage a 16-element number arrangement and determines if the sum of each row, column, main diagonal and nine 4-cell groups of consecutive numbers equals 34. If the sums equal 34, then the program initiates action to print an appropriate message along with the 16-element number arrangement.

The input data card that contains the number arrangement is illustrated in Figure 5-18.

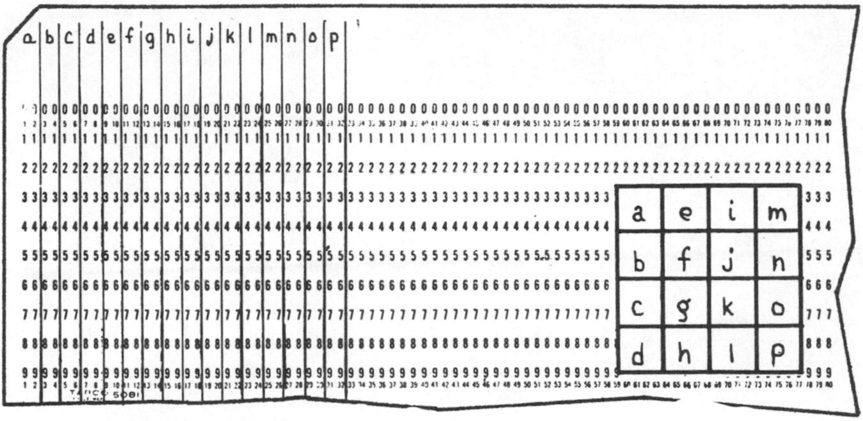

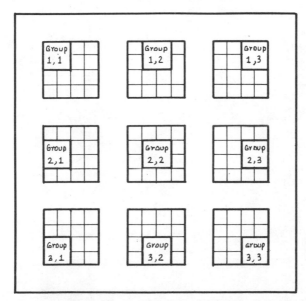

Each of the above letters may be represented by a 2-digit number. The relative position of each letter is shown in the square.

Figure 5-18. Input data card for the 34 puzzle checking program.

Figure 5-19 illustrates the method used by the program to identify the nine 4-cell groups of numbers. The first group is group 1,1, the second group is group 1,2, the third group 1,3, the fourth group 2,1, etc.

Figure 5-19. Nine 4-cell groups of the 34 puzzle.

A flowchart of the 34 Puzzle Checking Program is shown in Figure 5-20, and a FORTRAN program follows the flowchart.

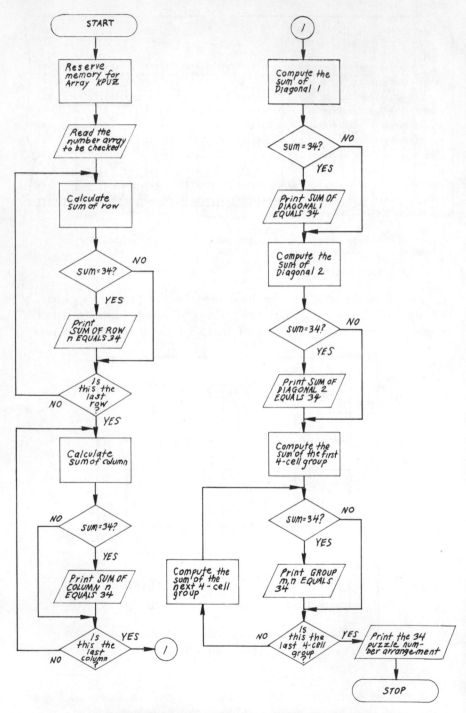

Figure 5-20. Flowchart of the 34 puzzle checking program.

```
C      34 PUZZLE CHECKING PROGRAM

       DIMENSION KPUZ(4,4)

C      READ ARRAY OF NUMBERS TO BE CHECKED

       READ 10,((KPUZ(K,L),K=1,4),L=1,4)

00010  FORMAT(16I2)

C      DETERMINE IF SUM OF EACH ROW EQUALS 34

       DO 50 I=1,4

       KSUM = 0

       DO 20 J=1,4

00020  KSUM = KSUM + KPUZ(I,J)

       IF(KSUM - 34)50,30,50

00030  PRINT 40,I

00040  FORMAT(12H0 SUM OF ROW,I3,10H EQUALS 34 )

00050  CONTINUE

C      DETERMINE IF SUM OF EACH COLUMN EQUALS 34

       DO 90 J=1,4

       KSUM = 0

       DO 60 I=1,4

00060  KSUM = KSUM + KPUZ(I,J)

       IF(KSUM - 34)90,70,90

00070  PRINT 80,J

00080  FORMAT(15H0 SUM OF COLUMN ,I3,10H EQJALS 34 )

00090  CONTINUE

C      DETERMINE IF SUM OF DIAGONAL 1 EQUALS 34

       KSUM = 0

       DO 100 I=1,4

       J=I

00100  KSUM = KSUM + KPUZ(I,J)
```

```
        IF(KSUM - 34)130,110,130
00110 PRINT 120
00120 FORMAT(28HOSUM OF DIAGONAL 1 EQUALS 34 )
C     DETERMINE IF SUM OF DIAGONAL 2 EQUALS 34
00130 KSUM = 0
      DO 140 I=1,4
      J = 4 - I + 1
00140 KSUM = KSUM + KPUZ(I,J)
      IF(KSUM - 34)170,150,170
00150 PRINT 160
00160 FORMAT(28HOSUM OF DIAGONAL 2 EQUALS 34 )
C     DETERMINE IF 4-CELL GROUPS EQUAL 34
00170 DO 220 L=1,3
      LL = L+1
      DO 220 M=1,3
      MM = M+1
      KSUM = 0
      DO 190 I=L,LL
      DO 190 J=M,MM
00190 KSUM = KSUM + KPUZ(I,J)
      IF(KSUM - 34)220,200,220
00200 PRINT 210,M,L
00210 FORMAT(7HOGROUP ,I2,2H ,,I2,15H  SUM EQUALS 34 )
00220 CONTINUE
C     PRINT  34 PUZZLE NUMBER ARRANGEMENT
      PRINT 230,((KPUZ(J,K),K=1,4),J=1,4)
00230 FORMAT(1H0,/,(4I5,//))
      STOP
      END
```

Program Output. The output illustrated in Figure 5-22 will be printed whenever the program reads the data card shown in Figure 5-21.

Figure 5-21. Input data card for the 34 puzzle checking program.

SUM OF ROW 1 EQUALS 34	GROUP 1 , 2 SUM EQUALS 34
SUM OF ROW 2 EQUALS 34	GROUP 2 , 2 SUM EQUALS 34
SUM OF ROW 3 EQUALS 34	GROUP 3 , 2 SUM EQUALS 34
SUM OF ROW 4 EQUALS 34	GROUP 1 , 3 SUM EQUALS 34
SUM OF COLUMN 1 EQUALS 34	GROUP 2 , 3 SUM EQUALS 34
SUM OF COLUMN 2 EQUALS 34	GROUP 3 , 3 SUM EQUALS 34
SUM OF COLUMN 3 EQUALS 34	
SUM OF COLUMN 4 EQUALS 34	

SUM OF DIAGONAL 1 EQUALS 34

SUM OF DIAGONAL 2 EQUALS 34

GROUP 1 , 1 SUM EQUALS 34

GROUP 2 , 1 SUM EQUALS 34

GROUP 3 , 1 SUM EQUALS 34

9	4	14	7
6	15	1	12
3	10	6	13
16	5	11	2

Figure 5-22. Printer output from the 34 puzzle checking program.

PART TWO

Games Proposed For Computer Solution

CHAPTER 6

Casino Games

6.1   Wheel of Fortune

THE WHEEL OF FORTUNE is a giant wheel with a diameter of about five feet, and is usually found at the front entrance of many casinos. The casino has a tremendous house percentage (over 18 percent) on this game, and if it were not for the tantalizing payoffs of 40 to 1, the game would probably not have many followers.

The rim of the wheel is divided into 50 sections. In 48 of these sections is paper money in denominations of $1, $2, $5, $10 and $20. The remaining two sections contain a Joker and a Flag. The Wheel of Fortune layout, which consists of seven corresponding numbers and symbols, is used by the players for placing bets.

The wheel is spun and players bet that it will come to rest with the pointer at a specified money denomination.

- A player will win even money if he bet on $1 and the pointer stopped at the $1 bill.
- A player will win $2 if he bet on $2 and the pointer stopped at the $2 bill.
- If the wheel stops at the $5 bill, the player will collect $5 if he bet on that value.
- If the wheel stops at the $10 bill, the player will win $10 if he has bet on that denomination.
- If the wheel stops at the $20 bill, the player will win $20 if he was betting on that value.
- The Joker and Flag pay off at 40 to 1 odds and a player betting on this value would collect $40 if the wheel stopped on either one.

On most wheels there are 22—$1 bills, 14—$2 bills, 7—$5 bills, 3—$10 bills, 2—$20 bills, 1—Joker and 1—Flag, located on the outer rim. Needless to say this game is one for the gambler to stay away from.

6.2   Roulette

Roulette is undoubtedly the most famous of all games of chance. It is perhaps the most glamorous of all casino games. Anyone visiting a casino, whether it be in Europe, Nevada, South America, Puerto Rico or Grand Bahama, is sure to find Roulette wheel action.* A very large part of Roulette's fascination lies in the color and beauty of the game. The green table cloth with bright gold, black and red markings as well as the highly polished mahogany and chrome roulette wheel produce an attractive spectacle when the wheel is spinning and chips and money are stacked on the layout. The plush casino surrounding certainly does not distract from this beauty.

Roulette is a simple game to learn how to play, yet it provides enough interest to make it extremely interesting. Roulette is played with a wheel which can be spun rapidly and a small ball which will eventually come to rest in one of the numbered compartments at the edge of the rotating wheel. This ball is caused to travel in a direction opposite the spinning wheel. As the ball loses speed it falls lower into the track. On the track are several metal diamond shaped studs which are present to deflect the ball. The ball finally becomes trapped in a number compartment which identifies the winning number. Bets may be made while the ball is revolving inside the track. The casinos do not pass up this extra chance for collecting additional bets.

Roulette openly advertises that the casino must take a *house percentage* in order to stay in business. Practically all casinos except those in Nevada use a wheel with 37 number compartments. This wheel, which is often referred to as the *European wheel*, has the numbers 1 to 36 and one *zero*. It is illustrated in Figure 6-1.

The American roulette wheel has the numbers 1 to 36, *zero* and a *double zero*. Figure 6-2 illustrates this wheel which is approximately 2 feet in diameter. The wheel is made of seasoned San Domingo mahogany, veneered and inlaid with rosewood and ebony. The 38 numbers are arranged around the wheel in such a manner that one might think there was no pattern. There is, however, a pattern: odd and even numbers alternate around the wheel; colors alternate, except red surrounds the green 00 and black surrounds the green 0. Zero and 00 are opposite numbers. Consecutive numbers appear opposite one another and succes-

---

* Monte Carlo operates simultaneously 33 double-layout single-zero roulette wheels.

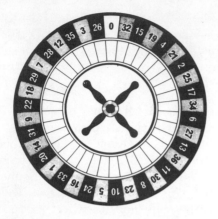

Figure 6-1. A single zero roulette wheel.

sive pairs of the same colored numbers (with two exceptions—0 and 00) total 37, i.e., $1 + 36 = 3 + 34 = 5 + 32 = 13 + 24 = 15 + 22 = 17 + 20$ etc.

The house percentage in the American roulette game is *5.26 percent* on all but one bet. The percentage may be figured in the following manner. Place a $1.00 chip on all 38 numbers on the layout. This means that you have a chip on all possible numbers on the wheel—the numbers 1 to 36, 0 and 00. Regardless of which compartment the ball falls in, the casino will pay off at 35 to 1 odds. This means that you will be paid $35.00 for the wagered $38.00. For this bet and all similar bets, the casino will relieve you of three chips out of every 38 you play. This results in a house percentage of 5.26 percent. The house has a slightly higher percentage on a bet placed on the combination of the numbers 0, 00, 1, 2, 3—7 34/38 percent.

Roulette has more attractive odds when using a single zero wheel. The house rakes in only *2.7 percent* of what is played.

In addition to the single zero wheel, Monte Carlo takes only half of what is bet on even-money bets. When a 0 appears, a bet on an even-

Figure 6-2. The *American* double zero roulette wheel.

money chance is neither won nor lost, but is placed in *prison*. The money remains in *prison* until a number other than 0 appears. The player will either win back his bet that is in *prison* or the house will collect the bet. The outcome will depend on whether this spin of the wheel was favorable or the opposite. The *prison* system reduces the odds against the player in such a manner that he will lose only 1.35 percent of what he bets on even-money bets. Is it a wonder that European roulette players call the American wheel a *money grabber?* Several years ago there was even an American wheel with *three house numbers—a zero, a double-zero* and an *eagle*. The remaining part of this section will only be concerned with the type of roulette wheel found in Nevada—the zero and double-zero wheel.

Bets are placed on a layout similar to the one shown in Figure 6-3. This layout is painted on a green cloth. The layout consists of a series of alternating black and red squares numbered 1 through 36. It is divided into three groups: first dozen, second dozen and third dozen. Another grouping is any number between 1 and 18 or between 19 and 36. There are spaces for betting Red or Black; Odd or Even. A player may also bet on numbers running horizontally: from 3 to 36, 2 to 35, 1 to 34. One may also place a wager on 0 and 00.

There are many ways to place bets at Roulette. *One number* or *straight bets,* where a chip or chips are placed on a single number, Red or Black, Odd or Even or any of the numbered groups. *Partial* bets, where a player's bet is split on a combination of numbers. By placing chips in various prescribed positions on the layout, a player can bet on one or more numbers or combinations of numbers. If the player wins, he will be paid off at the following odds:

• SINGLE NUMBERS PAY 35 TO 1

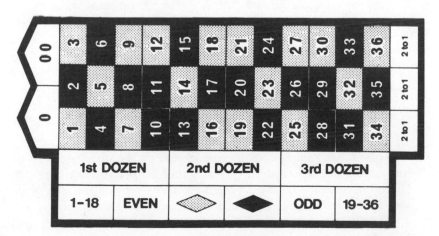

Figure 6-3. An American roulette table.

- DOUBLE NUMBERS PAY 17 TO 1
- THREE NUMBERS PAY 11 TO 1
- FOUR NUMBERS PAY 8 TO 1
- FIVE NUMBERS PAY 8 TO 1
- SIX NUMBERS PAY 5 TO 1
- COLUMNS PAY 2 TO 1
- HIGH OR LOW PAY EVEN
- RED OR BLACK PAY EVEN
- ODD OR EVEN PAY EVEN
- 1 TO 18 OR 19 TO 36 PAYS EVEN

The roulette bets are as follows:

1. A Single Number, or *Straight Bet*. A player may place a bet on any single number, including zero and double zero, by placing his chips on the chosen number. If that number appears, he will be paid 35 times the money he wagered.

2. A Two Number, or *Split Bet*. Any two numbers can be bet by placing chips on the line separating them, and if either of these numbers appear, the player will be paid 17 chips for each chip wagered.

3. A Three Number, or *Street Bet*. A bet may be placed on the line to the right or left of any row or *street* of the three numbers running across the layout. If any of the three numbers comes up, the player will be paid 11 times his original stake. The following three number combinations may be made by placing the chip so that it partially lies in each number square: 0-1-2, 00-2-3, 0-00-2.

4. A Four Number, or *Square Bet*. Any four numbers may be bet by placing a chip or chips at the intersection point of any four numbers on the layout. The winning payoff is 8 to 1.

5. A Five Number, or *House Special Bet*. There is only one way to bet on five numbers by the placement of one chip. This bet is on the numbers 0, 00, 1, 2 and 3 and is made by placing a chip on the line that separates the 0, 00 and the 1, 2, 3. The payoff of this bet is 6 to 1; however, the house percentage is 7 34/38 percent on this bet. Needless to say that this is the poorest of all Roulette bets.

6. A Six Number, or *Line Bet*. A chip placed on the intersection of the line separating any two rows and either outside vertical line constitutes a bet made on the six numbers of those two rows. The player will collect five chips for each one bet on a winning number.

7. A *Column Bet*. The Roulette layout has three columns of 12 numbers. They are composed of the numbers

Column 1—1,4,7,10,13,16,19,22,25,28,31,34
Column 2—2,5,8,11,14,17,20,23,26,29,32,35
Column 3—3,6,9,12,15,18,21,24,27,30,33,36

A column bet is made by placing a chip or chips in the space at the bottom of the selected column. If any one of the numbers comes up, the player will collect at 2 to 1 odds.

8. A *Dozen Bet.* The layout is divided into three squares of 12 numbers: 1st twelve (or 1st dozen), 2nd twelve (or 2nd dozen), 3rd twelve (or 3rd dozen). A bet placed in one of the squares labeled first 12, second 12, or third 12 will be paid at 2 to 1 odds whenever a number of that 12 comes up.

9. *High-Low Bets.* A bet placed in the area marked 1-18 will win if a number in that range appears. Likewise a bet placed on 19-36 would win only if the number lies in the range from 19 to 36. This bet pays one to one, even money.

10. *Red-Black Bets.* A bet placed on Red will win if one of the following numbers comes up:

1, 3, 5, 7, 9, 12, 14, 16, 18, 19, 21,
23, 25, 27, 30, 32, 34, 36

A wager placed on Black will win if any one of the following numbers appears:

2, 4, 6, 8, 10, 11, 13, 15, 17, 20,
22, 24, 26, 28, 29, 31, 33, 35

This is an even money bet.

11. *Even-Odd Bet.* A bet placed on Even will win if any of the following numbers comes up:

2, 4, 6, 8, 10, 12, 14, 16, 18, 20,
22, 24, 26, 28, 30, 32, 34, 36

Likewise, a bet placed on Odd will win only if an odd number appears:

1, 3, 5, 7, 9, 11, 13, 15, 17, 19,
21, 23, 25, 27, 29, 31, 33, 35

All even-odd bets will be paid off at one to one odds.

6.3 Keno

Chinese immigrants, brought to this country for railroad construction, introduced a new game to the West. Today, virtually the same game is played in modern casinos under the name of *Keno*—yet its origin reaches far back into Chinese history.

The original Chinese game used 80 discs, each inscribed with a unique Chinese character. These characters were from an ancient and famous poem of luck. The discs were mixed in a bowl and then placed in four smaller bowls. Each of the smaller bowls contained 20 discs. The discs were later replaced with wooden balls with characters painted on them. In the early 1900's, a metal cage was used instead of the bowl to mix up the balls. The balls were rather noisy and were later replaced with a good quality ping pong ball. Most of today's modern casinos use an electronic blower to mix up the balls; however, many gambling establishments still use the metal cage.

Keno was introduced in Nevada shortly after 1930. It was called Racehorse Keno at that time because the names of horses were used on the balls instead of Chinese characters. The horses' names were later replaced with the numbers 1 through 80. The modern game of Keno, using numbered balls and mechanical and electronic equipment, still shows some trace of its Chinese origin. An oriental brush and ink are used to mark numbers on Keno tickets.*

A Keno ticket is illustrated in Figure 6-4. It is marked off in 80 squares and is further subdivided into two rectangles, with the numbers one to 40 in the upper rectangle and 41 to 80 in the lower. This separation was very important to the Chinese. The upper area was called *Yin* and the lower *Yang*. It was the separation of Night and Day—Heaven and Earth—Sun and Moon—Male and Female and other opposites. They were significant because an event of importance occurring in the night would influence a player to play in the first 40 squares (Yin), whereas a flood or other disaster would influence him to play in the lower area (Yang).

Keno is a group game where many players are active in the game at the same time. Most casinos featuring Keno have many games throughout a 24 hour period. Using the casino-provided brush/ink or crayon marker, a player marks anywhere from one up to 15 spots (numbers) on a Keno ticket. Figure 6-5 illustrates a sample marking of a 10-spot Keno ticket. After the player marks a ticket, he hands it to a Keno writer with the amount he wishes to play. The writer makes a copy Keno ticket as a receipt.

Numbers are then selected either by an automatic blower or from a wired cage. As each ball is selected, its number is called and flashed on an electrically lit board which can represent the 80 numbers on the Keno ticket. When the 20th number has been selected, the game ends. If you were lucky enough to choose the proper number of numbers that were selected, you win.

---

* Some casinos have replaced the brush and ink with a black crayon marker.

PLAY
KENO

Figure 6-4. Keno ticket.

PLAY KENO
YOU CAN WIN $25,000.00

Figure 6-5. Keno ticket with the marked spots (numbers 10, 13, 20, 21, 25, 34, 45, 57, 62 and 68).

Keno plays are referred to as *1-spot* where one number is played, *2-spots* where two numbers are played, *3-spots* where three numbers are played, *4-spots* where four numbers are played and so on up to *15-spots* where 15 numbers are played. Bets will be paid off in accordance with the tables provided by the casino. A Reno, Nevada casino Keno payout table is shown in Table 6-1. For example, if a player played $1.00 on a *1-spot* ticket and his number showed, he would be payed $3.20. If a player bets $1.00 on a *5-spot* ticket, and the five numbers just happened to come up, he would be paid $332. Don't count on this happening too often as the odds against it are greater than 1 in 1550. It is possible to win $25,000 on a 50¢ *14-spot* ticket. Of course the chance is rather remote: 1 chance in 38,910,016,281.

There are other Keno bets that are not shown in Table 6-1. These bets involve selecting groups of numbers and are usually described in Keno pamphlets that are provided by the casinos.

Everyone likes to win at games of chance and Keno is no exception. Many players feel that they have better luck if following some system. An old Keno system states *You may chase the Old Man of China,* which means that the player is to mark different numbers each game, hoping to catch up with the numbers that have been coming up. Another system says *Let the Old Man catch you,* on the premise that sooner or later all 80 numbers must come up, and if you use the same numbers on all your tickets, you will hit a big winner. Keno has a house percentage of greater than 20% and will relieve a player of his bets at an extremely fast rate whatever system he uses. Admittedly, Keno's great attraction lies in the fact that a player can win an astronomical amount of money in return for a small bet. The player should never forget that the odds against this happening are more astronomical than the amount of money he could ever win. For example, the odds against the player picking 15

Table 6-1. Keno Payoff Table.

1   MARK 1 SPOT

Catch	Play $ .50	Play $1.00
1 - Pays	$ 1.60	$ 3.20

2   MARK 2 SPOTS

Catch	Play $ .50	Play $1.00
2 - Pays	$ 6.50	$ 13.00

3   MARK 3 SPOTS

Catch	Play $ .50	Play $1.00
2 - Pays	Money Back	Money Back
3 - Pays	$ 23.50	$ 47.00

4   MARK 4 SPOTS

Catch	Play $ .50	Play $1.00
2 - Pays	Money Back	Money Back
3 - Pays	$ 2.50	$ 5.00
4 - Pays	59.00	118.00

5   MARK 5 SPOTS

Catch	50c Tkt.	$1.00 Tkt.	$2.00 Tkt.
3 - Pays	$ 1.50	$ 3.00	$ 6.00
4 - Pays	13.00	26.00	52.00
5 - Pays	166.00	332.00	664.00

6   MARK 6 SPOTS

Catch	50c Tkt.	$1.00 Tkt.	$2.00 Tkt.
3 - Pays	$ .50	$ 1.00	$ 2.00
4 - Pays	2.80	5.60	11.20
5 - Pays	55.00	110.00	220.00
6 - Pays	620.00	1,240.00	2,480.00

7   MARK 7 SPOTS

Catch	50c Tkt.	$1.00 Tkt.	$1.50 Tkt.
0 - Pays	$ 1.00	$ 2.00	$ 3.00
4 - Pays	1.00	2.00	3.00
5 - Pays	7.00	14.00	21.00
6 - Pays	150.00	300.00	450.00
7 - Pays	1,600.00	3,200.00	4,800.00

8   MARK 8 SPOTS

Catch	50c Tkt.	$1.00 Tkt.	$3.00 Tkt.
4 - Pays	$ .50	$ 1.00	$ 3.00
5 - Pays	7.50	15.00	45.00
6 - Pays	55.00	110.00	330.00
7 - Pays	447.50	895.00	2,685.00
8 - Pays	2,250.00	4,500.00	13,500.00

9   MARK 9 SPOTS

Catch	50c Tkt.	$1.00 Tkt.	$4.00 Tkt.
4 - Pays	$ .20	$ .40	$ 1.60
5 - Pays	2.55	5.10	20.40
6 - Pays	25.40	50.80	203.20
7 - Pays	158.10	316.20	1,264.80
8 - Pays	1,428.55	2,857.10	11,428.40
9 - Pays	3,214.20	6,428.40	25,000.00

10   MARK 10 SPOTS

Catch	Play $ .50	Play $1.00
5 - Pays	$ 1.00	$ 2.00
6 - Pays	9.00	18.00
7 - Pays	90.00	180.00
8 - Pays	650.00	1,300.00
9 - Pays	1,300.00	2,600.00
10 - Pays	5,000.00	10,000.00

11   MARK 11 SPOTS

Catch	Play $ .50	Play $1.00
5 - Pays	$ .50	$ 1.00
6 - Pays	5.00	10.00
7 - Pays	38.00	76.00
8 - Pays	240.00	480.00
9 - Pays	800.00	1,600.00
10 - Pays	2,000.00	4,000.00
11 - Pays	5,000.00	10,000.00

12   MARK 12 SPOTS

Catch	Play $ .50	Play $1.00
5 - Pays	$ .30	$ .60
6 - Pays	2.60	5.20
7 - Pays	18.70	37.40
8 - Pays	106.50	213.00
9 - Pays	374.00	748.00
10 - Pays	912.80	1,825.60
11 - Pays	1,916.60	3,833.20
12 - Pays	5,000.00	10,000.00

13   MARK 13 SPOTS

Catch	Play $ .50	Play $1.00
7 - Pays	$ 9.00	$ 18.00
8 - Pays	53.00	106.00
9 - Pays	460.00	920.00
10 - Pays	2,200.00	4,400.00
11 - Pays	4,240.00	8,480.00
12 - Pays	6,000.00	12,000.00
13 - Pays	8,000.00	16,000.00

14   MARK 14 SPOTS

Catch	Play $ .50	Play $1.00
7 - Pays	$ 5.00	$ 10.00
8 - Pays	28.50	57.00
9 - Pays	197.00	394.00
10 - Pays	700.00	1,400.00
11 - Pays	4,000.00	8,000.00
12 - Pays	9,000.00	18,000.00
13 - Pays	18,500.00	25,000.00
14 - Pays	25,000.00	25,000.00

15   MARK 15 SPOTS

Catch	Play $ .50	Play $1.00
7 - Pays	$ 3.90	$ 7.80
8 - Pays	14.00	28.00
9 - Pays	82.00	164.00
10 - Pays	315.00	630.00
11 - Pays	1,300.00	2,600.00
12 - Pays	6,000.00	12,000.00
13 - Pays	14,000.00	25,000.00
14 - Pays	25,000.00	25,000.00
15 - Pays	25,000.00	25,000.00

spots on a 15-spot ticket is greater than 1 in 428,010,179,098 chances. However, the maximum casino payoff is limited to $25,000.

6.4 Slot Machines

Most gambling devices have a very old history, usually originating in Europe or the Orient. The slot machine is a device that is all-American. The first slot machine was the *Liberty Bell** which was invented by

---

* The Liberty Bell may be seen in the collection of old slot machines at the Liberty Bell Saloon and Restaurant in Reno, Nevada.

Charlie Fey of San Francisco in 1895. This machine had three wheels
bearing symbols of bells, horseshoes, hearts, diamonds, spades and a star.
A player would insert a nickel and pull the lever to start the wheels
spinning. If one of the following combinations of symbols appeared on
the payline, the player was rewarded with a payoff in nickels.*

  ⚲ ⚲ ⚲   10 drinks

  ♡ ♡ ♡   8 drinks

  ◊ ◊ ◊   6 drinks

  ♤ ♤ ♤   4 drinks

  ∩ ∩ ☆   2 drinks

                      1 drink

  ∩ ∩ ▭

Charlie Fey's twenty dollar machine was an immediate success. He
refused to sell or lease manufacturing rights to his slot machine which
resulted in H. S. Mills, a Chicago manufacturer of gaming machines,
developing a much nicer machine. The Mills machine was called the
Operators Bell and featured symbols of cherries, plums, bells, lemons,
and bars. Mills' machine was also extremely popular, and it was just a
matter of a few years until the slot machine appeared in bars and
gambling houses throughout the country. Today, the slot machine may
also be found in England, Puerto Rico, Japan, Grand Bahama, Domin-
ican Republic and several states in the United States.

As nearly everyone knows, the slot machine is a mechanical device
that will absorb coins just about as fast as you can feed it. After the
player inserts a coin into the machine, he may pull a handle or lever that
starts three independent reels spinning. If the reels stop with certain
symbols appearing in the pay-line, the machine dispenses a specified
number of coins to the player. The machines may be set to pay off a
certain percentage of all money put into the machine.

There are twenty symbols on each vertical wheel of a slot machine.
The number of different combinations that can occur on a three wheel
machine is 20 x 20 x 20 or 8000 possible combinations. Of course, the
odds on winning a jackpot are not 8000 to 1, as intermediate payoffs of
smaller amounts, such as a five-coin payoff for two cherries or a ten-coin
payoff for three oranges, tends to lower the odds. Using a liberal paying
machine as an example and assuming that all intermediate coin payoffs
were put back into the machine, we may find the odds against a player
winning the jackpot around 400 to 1.

---

* Although the pay-off table specifies drinks, the slot machine paid off in nickels.

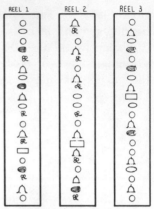

Figure 6-6. The Programmers Bell.

Figure 6-7. Symbol configuration of the wheels in the Programmers Bell.

ßℛ	anything	anything	3 coins
ßℛ	ßℛ	anything	5 coins
◯	◯	BAR	6 coins
△	△	◯	8 coins
⬭	⬭	◯	10 coins
▱	▱	▭	15 coins
◯	◯	◯	18 coins
▱	▱	▱	20 coins
BAR	BAR	BAR	JACKPOT 200 coins

Figure 6-8. Payoff table for the Programmers Bell.

There are many different kinds of slot machines. Many Nevada machines have several different jackpot payoffs. Others have only one jackpot payoff. At times, two, three and four slot machines are bolted together and operated by one lever. Jackpots on these machines are usually several thousand dollars. What are the odds of hitting a jackpot on a *four-machine* combination? Only one chance in 4,096,000,000,000,-000. Another machine that is rapidly becoming popular in Nevada casinos is the *four-reel machine*.

Consider the slot machine shown in Figure 6-6—*The Programmers Bell.* Figure 6-7 illustrates the three wheels of this machine and the 20 symbols on each wheel.

The Programmers Bell payoff table is shown in Figure 6-8.

The combination of one cherry on reel 1 and any other symbols on reels 2 and 3 results in a payoff of three coins. Cherries appearing on reels 1 and 2 with any other symbol on reel 3 will payoff five coins. Other payoff values are shown in the payoff table.

Table 6-2 lists the total number of times a specific symbol appears on the reels.

Table 6-2. Breakdown of Symbols on The Programmers Bell.

	Reel 1	Reel 2	Reel 3
Cherries	4	6	0
Oranges	5	4	7
Bells	4	6	5
Lemons	3	2	4
Watermelons	3	1	3
Bars	1	1	1

A method of computing the favorable percentage for The Programmers Bell is shown in Table 6-3. This table assumes a payoff for 8000 plays.

Table 6-3. Coin Payoff Table for The Programmers Bell.

Symbol Configuration	Coin Payoff	Number of Possible Ways		Total Coin Payoff
🍒 — —	3	4 x 10 x 10	400	1200
🍒 🍒 —	5	4 x 6 x 10	240	1200
O O BAR	6	5 x 4 x 1	20	120
△ △ O	8	4 x 6 x 7	168	1344
O O O	10	3 x 2 x 4	24	240
⊘ ⊘ BAR	15	3 x 1 x 1	3	45
O O O	18	5 x 4 x 7	140	2520
⊘ ⊘ ⊘	20	3 x 1 x 3	9	180
BAR BAR BAR	200	1 x 1 x 1	1	200
				7049

The total coin payoff of 7049 coins represents an average return of coins for each 8000 coins put into the machine. The net loss is 951 coins; thus, the favorable percentage is 951/8000, or 11.89 percent. This percentage figure exists only for the machine configuration being discussed. Casino operators are reluctant to release the house percentage of their slot machines. It has been estimated by a variety of gambling experts as being somewhere between 50 and 3 percent. Perhaps a good average might be a percentage near the sample machine—11 or 12 percent. Of course the house percentage of slot machines will vary depending upon where the machines are located. It is an established fact that slot machines in the Las Vegas, Nevada downtown casinos are much more liberal in payoffs than the machines found in the plush hotel casinos located on the strip. The slot machine is without a doubt the most consistent money-making device in the Nevada casinos. Some 23,000 machines are now operating in Nevada. A typical machine might *win* around $5,000 a year.

In order to detect any slot machine that might become too liberal with the house's money, digital computers are being employed to monitor the payoff. The computer keeps account of the amount of money that is put into the machine, the amount paid out, and various other accounting percentages. This information is printed for each machine. It is easy to see how the performance of a machine is determined with this information being available.

6.5 Chuck-A-Luck

Chuck-A-Luck, sometimes called *Bird Cage* or *Hazard,* is a casino banking game. The game is not as popular as Roulette, Craps or Blackjack because the house percentage against them is less.

Chuck-A-Luck is a game where three dice are located in an hour-glass shaped wire cage. The three captive dice tumble from one end of the cage to the other as the cage is spun and finally come to rest at one end of the cage.

Some casinos use a very simple layout consisting of six numbers. On this layout, players may bet only on the single numbers 1, 2, 3, 4, 5 or 6. After the cage is spun and the three dice come to rest, the house takes the bet if the number does not appear in the three dice. If it appears once, the house pays even money; twice, the bank pays 2 to 1; 3 times, the bank pays 3 to 1. The house expects to win $17 of every $216 that is bet. This gives the house an advantage of about 8 percent.

Figure 6-9 illustrates a more elaborate Chuck-A-Luck layout. As indicated in this layout, the house not only accepts bets on single numbers, but on odd, even, high, low, triplets and point totals of three dice as well. The player may make the following bets:

Low Bet: A bet that the total count of the three dice will be between 4 and 10.

High Bet: A bet that the total count of the three dice will be between 11 and 17.

Odd Bet: A bet that the total count of the three dice will be an odd value: 5, 7, 9, 11, 13, 15, 17.

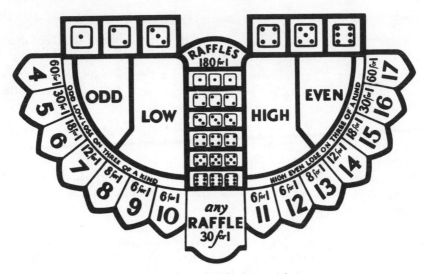

Figure 6-9. Elaborate Chuck-A-Luck layout.

Even Bet: A bet that the total count of the three dice will be an even number: 4, 6, 8, 10, 12, 14, 16.

Raffle: A bet that a *specific* triplet or three-of-a-kind will appear on the three dice.

Table 6-4. Chuck-A-Luck Odds.

Bet	True Odds	House Odds
Low	111 to 105	Even
High	111 to 105	Even
Odd	111 to 105	Even
Even	111 to 105	Even
Raffle	215 to 1	179 to 1
Any Raffle	35 to 1	29 to 1
4	71 to 1	59 to 1
5	35 to 1	29 to 1
6	23 to 1	17 to 1
7	67 to 5	13 to 1
8	65 to 7	7 to 1
9	8 to 1	5 to 1
10	7 to 1	5 to 1
11	7 to 1	5 to 1
12	8 to 1	5 to 1
13	65 to 7	7 to 1
14	67 to 5	13 to 1
15	23 to 1	17 to 1
16	35 to 1	29 to 1
17	71 to 1	59 to 1

Any Raffle: A bet that *any* triplet or three-of-a-kind will appear.

Number Bets: The player may bet on the total count of the three dice: 4, 5, 6, 7, 8, 9, 10, 11, 12, 13, 14, 15, 16, 17.

One Number: The player may still place a bet on a single number as was done in the simple Chuck-A-Luck layout. In some casinos the 3 to 1 bet is lost whenever a triplet appears.

Whenever three-of-a-kind appears, the house collects all bets on High-Low, Even-Odd and combination number bets. For example, a player betting on the number 12, Even and High would lose if the dice total was 4-4-4. Table 6-4 shows the correct odds and the odds offered by the house. The reader should note that the Chuck-A-Luck layout disguises the true odds by using the word *for*. For example, if a player won a bet on a 6 for 1 layout position, he would receive his $1 back together with $5, for a total of $6. This is equivalent to odds of 5 to 1. The best Chuck-A-Luck bet is on the Even, Odd, High, Low bets. The house collects less than three cents per dollar bet. On all other Chuck-A-Luck bets the house will collect from 9 to 22 cents of each dollar wagered.

6.6   Baccarat and Chemin de Fer

Baccarat is a popular card game in European casinos; however, it is relatively new in American casinos. The game dates back to 1500 where it was played in Italy. Baccarat was introduced in Nevada casinos in the early 1920's and even today is played only in a few plush hotel casinos.

Chemin de Fer, a similar game, is also starting to become popular in the Nevada casinos. The primary difference between Chemin de Fer and Baccarat is that in the latter game the bank remains with one banker* whereas Chemin de Fer passes the bank around the players. The Baccarat bank goes to the highest bidder—the player who agrees to put up the largest bank.

Besides the banker, ten other players may be *active* in the game. Six packs of 52 cards each are well shuffled together and used as one.** The cards are dealt from an open-topped, open-ended box called a *shoe*.

The banker places his bank or stake before him on the table. The other players take their seats, even number of players on each side of the banker. A player may bet against the whole of the banker's stake by

---

* In many casinos the house is the bank.

** Three, four or five, seven and eight decks of cards have been known to be used in dealing shoes.

calling *banco*. However, if no one individual cares to take the whole stake, the other players may combine their bets to match the amount of the banker's stake. The player having the largest amount in this takes it upon himself to represent the other players against the bank. This player is called the *active player*.

The object of either game is to achieve a total of 9, *La Grande* (the big one), with two or three cards. In the event that 9 doesn't appear, the next highest hand is a total of 8, *La Petite* (the little one). The next highest count is 7, then 6, etc.

Baccarat and Chemin de Fer are modulo 10 games where a total of card values may never exceed nine. * Face cards and tens count zero, aces count one and all other cards count their index value. The following examples should make the counting scheme appear clearer.

Cards	Count
5-4	nine
J-5	five
5-4-Ace	zero
2-9-3	four
7-6	three
K-J	zero
Ace-K	one

In Baccarat, the banker deals one card each to the players on his right and left and one to himself, then repeats this dealing operation for another round of cards. The player on the right plays the hand dealt to his side, and the player on the left plays the hand dealt to his side. The banker always plays his own hand. When a player has an eight or nine and the banker does not have an equal number, the player may turn his cards over and win. If the banker has an eight or nine, however, and neither player has such a value, the banker wins everything on the table. If no one has eight or nine, the banker offers a third card to the player on his right, who may either take it or refuse it. It is then offered on the left. If both players refuse it, the dealer must take it himself, but if either player accepts it, the banker is not obligated to take one himself. The cards dealt are left face up on a layout similar to the one shown in Figure 6-10. All cards are now exposed and the banker pays the players who are nearer nine than himself and collects from the players if his total is closer to nine.

In Chemin de Fer, the bankers deals one hand to himself and another hand to the *active player*. If either the active player or the banker has a *natural* (eight or nine), it is shown immediately and the bet is settled

---

* Baccarat means *nothing*. Thus a player holding nothing but face cards or a 7 and a 3 is baccarat.

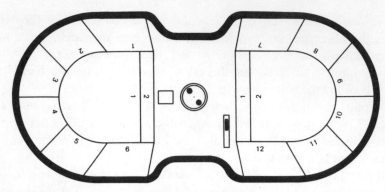

Figure 6-10. A baccarat layout.

as in Baccarat. If neither the player nor the banker had a natural, the player, and then the banker, may stand or may draw one card. The hands are exposed and if the banker wins or ties, he remains the banker; otherwise, the next player in a counter clockwise direction becomes the banker. Of course, a player may always pass the bank. Good players find it a bad policy not to draw with five or less in their first cards or to draw with more than five. Table 6-5 illustrates the players' strategy.

Table 6-5. Players Strategy in Baccarat and Chemin de Fer. Face Cards Count 10 or Nothing and Aces Count 1.

Total Count of Player's First Two Cards	Recommended Action for the Player to Follow
1 - 2 - 3 - 4 - 5 - 10	Draw a Card
6 - 7	Stand
8 - 9	Turn Cards Over

In Baccarat there is little use of skillful play, for the play in most situations is clearly laid out by rules. The banker's play is determined by the card given to a player. Thus, as shown in Table 6-6, the banker's action depends on the player's drawn card. For example, if the banker holds a total card count of 3, and if the player draws either a 1 (Ace), 2, 3, 4, 5, 6, 7 or 10 (face card or 10), the banker *must* draw a card. If the player draws an 8, the banker *cannot* draw a card.

Baccarat and Chemin de Fer are games for the wealthy as the minimum bet in most casinos is $5.00; the usual bet is $20.00 and many games are played with stacks of $100.00 bills or $500.00 chips. By consulting with the casino management it may even be possible to have the maximum bet raised to $1500.00. The edge against the player at either Baccarat or Chemin de Fer is slightly more than 1 percent. The games, like

Table 6-6. Dealer's Strategy in Baccarat and Chemin de Fer. Face Cards Count 10 or Nothing and Aces Count 1.

Value Of Banker's Hand	Banker Draws When Giving Player	Banker Does Not Draw When Giving Player
1	Always Draw	
2	Always Draw	
3	1 - 2 - 3 - 4 - 5 - 6 - 7 - 9 - 10	8
4	2 - 3 - 4 - 5 - 6 - 7	1 - 8 - 9 - 10
5	4 - 5 - 6 - 7	1 - 2 - 3 - 8 - 9 - 10
6	6 - 7	1 - 2 - 3 - 4 - 5 - 8 - 9 - 10
7	Stand	
8 - 9	Turn Cards Over	
10	Always Draw	

roulette and craps, are games of pure chance, and even though the percentage against the player is low, his bankroll will eventually be taken by the house if it is left on the Baccarat table.

6.7 Faro

*Faro,** the aristocrat of the gambling games, is perhaps the oldest banking game in the world. It was an extremely popular game in European gaming rooms during the 17th century. Faro was also the principal game in American gambling houses up until the 20th century. During the 20th century the popularity of Faro decreased as the game of Craps became the American favorite. By 1940, a Faro game could hardly be found outside a Nevada gambling casino and even today there are only a few casinos in Las Vegas and Reno that offer the game.

Faro is a banking game which any number of players can play. Fifty-two playing cards are shuffled and placed in a dealing box. This box has a spring pressing the cards upward and a slit permitting only one card to be removed at a time. Two of the cards in the box are never in play. One is the first card in sight, on the top of the box, and the other is the last card in the deck. The first card is called *soda* and the last the *hockelty* or *hoc*.

After the first card is withdrawn from the box, it is placed on the dealer's right, a little way from the box. The next card to come out of the box is called a *loser* and is laid close to the right side of the box. The card which is now in sight on the top of the box is the *winner* for that turn. (See Figure 6-11.)

Every card must either win or lose, except the soda and the hoc. The

---

* Faro is not of American origin. It was originally called *Pharaoh*.

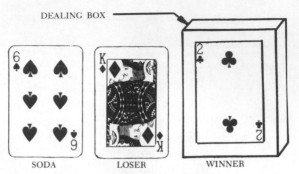

Figure 6-11. Placement of cards in Faro.

winning cards and the losing cards are kept in separate piles. The winners are placed on the soda card.

Figure 6-12 illustrates a Faro layout which consists of a complete suit of spades, enameled on a green table cloth, with enough space between the cards to allow bets to be played.

All bets are placed against the house. If a player thinks that a card of any denomination, such as a six, will *win* the next time it appears, he places his bet on the six of spades on the layout. If he thinks the next six that shows will *lose,* he places a wooden counter or a *copper* on the top of his bet. This is called *coppering the bet* and was named because copper coins were used originally for this purpose.

No *action* can be had on a bet until the card bet upon appears. If it does not appear after a turn has been made, the player is at liberty to

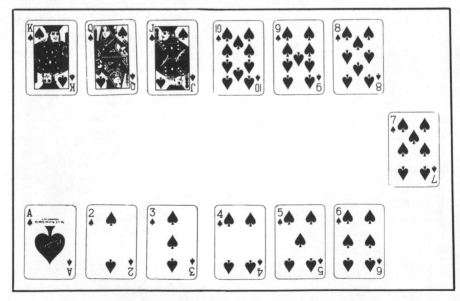

Figure 6-12. Faro layout.

change his bet, or to, remove it altogether. Each bet is made for the turn only, unless the player chooses to leave it until he gets some action on it.

If the winning and losing cards are both of the same denomination, it is called a *split,* and the dealer takes half of the chips bet on them to win or to lose.

Bets may be placed as to take in more than one card. Thus, a chip placed between the King and the 2, will play both cards and one between the 9 and 4 will play both cards. A chip between 6, 7 and 8 will play all three cards. A bet placed in any of the squares inside the layout, such as diagonally between the 4, J, 3 and 10, plays all four cards. A chip placed on the *outside* corner of a card means that the player is playing that card and the next-out-one card to it, in the direction indicated by the corner the bet is placed on. Thus, a chip placed on the outside corner of the 5 in the direction of the Ace plays the 5 and the 3. A bet behind a card on the outer edge takes in three cards. One outside the 4 would take in the 3, 4, 5. When a bet is playing more than one card, the bet is decided by whichever of the cards shows first.

After each turn, the banker first picks up all the bets he wins, and then pays all he loses. The dealer always has a *casekeeper* to assist him, who keeps a record of all cards as they come out of the box, by slipping beads along wires on an apparatus that resembles a Chinese abacus, as seen in Figure 6-13. There are four beads for each card. This counting

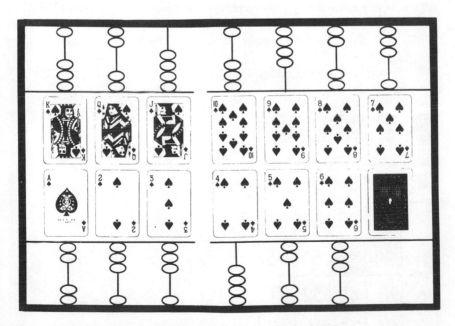

Figure 6-13. Faro counting board.

board is used to prevent players from betting on *dead* cards, and also to let them know how many of any denomination are still to come.

On the *last* turn, if three different cards are in the box, any player who can guess the sequence will be paid 4 for 1. Thus, if the Jack, 3, King are left in the box, you guess the order of King, Jack, 3, you place your chips on the side of the King facing the Jack. If you guessed correctly, you win. The percentage against *calling the turn* is 16⅔ percent.

If there are two cards of one denomination in the last turn, it is called a *cat-hop*, and any player who can call it correctly will be paid 2 for 1.

6.8   Bingo

Bingo, the casual American gambling game, is popular in both gambling casinos and Aunt Beth's gambling parlor. Ninety percent of all Bingo players are the fairer sex. Charitable organizations, social clubs, and several churches use Bingo as a regular fund-raising method. The game of Bingo is relatively new to America, being introduced at the end of World War I.

There are many variations of Bingo. The variations described in this section are those generally played in Nevada gambling houses.

A 6 by 4 inch Bingo card is made from lightweight cardboard and has a design that is divided into twenty-five squares as seen in Figure 6-14. The letters B-I-N-G-O appear at the headings of the card columns. The card contains a free play in the center square and is usually marked as such. The other 24 squares on the card contain numbers in the range 1 through 75. Bingo card columns contain numbers in the following ranges:

Column under B .... 1 through 15
Column under I .... 16 through 30

B I N G O

(1 TO 15)   (16 TO 30)   (31 TO 45)   (46 TO 60)   (61 TO 75)

4	30	31	58	62
7	24	39	57	69
11	28	FREE SPACE	50	66
6	18	41	52	71
9	26	35	46	67

Figure 6-14. Bingo card.

Column under N .... 31 through 45
Column under G .... 46 through 60
Column under O .... 61 through 75

A complete set of Bingo cards contains over 3000 cards with different numbers.

The object of the basic game is to get five numbers in a row, column or diagonal. The center square may be counted as a number that has been called. A *caller* draws the numbers, which are painted on ping pong balls, from an air blower or a wire cage. As the balls or numbers are selected, the caller announces the number and illuminates the number on a large viewing board.

A player marks his card or cards with beans, coins, or any other type of small markers as the numbers are drawn. As soon as a player has

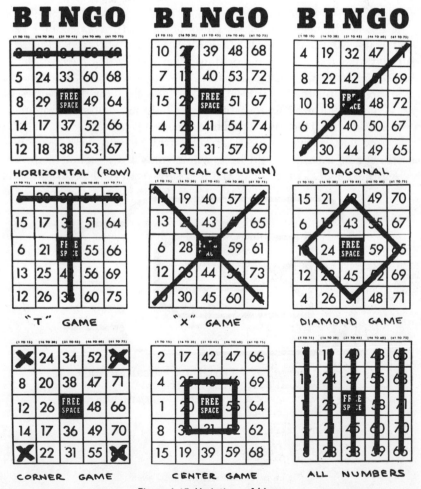

Figure 6-15. Variations of bingo.

covered a specified pattern, he calls *Bingo.* After the numbers on the card are checked against the master board, and they agree, the game is over.

There are many variations of Bingo. A few ways of marking a card are shown in Figure 6-15.

6.9   Trente et Quarante (Rouge et Noir)

Trente et Quarante is a banking game that is played in Monte Carlo and other French casinos. It is not played in American casinos.

The Trente et Quarante Layout, Figure 6-16, has a triangle on one end marked *inverse* and in the middle a square marked *couleur* (color). On each side of the square are two diamond shaped places, one *rouge* (red) and the other *noir* (black). Before the game starts, players may place bets on these four divisions.

The *dealer* shuffles together six packs of 52 cards. Aces count 1 each, K-Q-J count 10 each, and all other cards count their face value. The dealer lays out a row of cards, dealing them one by one, face up on the table, announcing the total count until the count reaches 31 or greater. It is impossible for the count to exceed 40. This first row dealt represents noir. Having dealt the first row, the dealer lays out another row of cards in the same way. This row represents rouge.

The row with a cumulative total nearest to 31 wins. Had a bet been placed on noir then, it would be a winner only if the count of the first row was nearer 31. The color of the first card dealt is of the color desig-

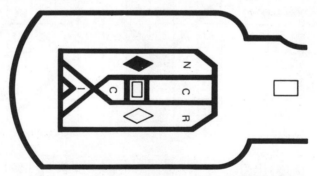

C- Bet on <u>couleur</u> (that the first card in each
    row will be the color of the row)
I- Bet on inverse (that the first card will be
    the opposite colour)
N- Noir (black)
R- Rouge (red)

Figure 6-16. Trente et quarante layout.

nating the winning row; all bets on *couleur* are paid. If this card is of the opposite color, *inverse* wins.

When both rows have the same total count and the count is 32 or greater, all bets are a stand-off. This is called a *refait*. If exactly 31 is dealt in each row, the dealer takes half of all money on the table, or players have the option of putting all bets *in prison*, which means that bets remain there until the next deal. A player withdraws his bet if the same couleur he bet upon comes up this time; otherwise he loses the entire bet.

The house's advantage is represented by a *refait* at 31. It has been estimated by various experts that a refait at 31 occurs once in every 40 deals.

6.10   Craps

Craps is undoubtedly the game that attracts the American gambler. Crap tables are found in all the Nevada gaming casinos as well as those found in Puerto Rico, Monte Carlo and the Bahamas. The game of Craps developed from an old English game called Hazard.

Craps is played by tossing two dice on a layout similar to the one illustrated in Figure 6-17. This layout is marked on a table with high retaining walls. Bets are also placed on this layout. The dice are thrown so as to hit the retaining wall and bounce back on the layout. All bets are placed against the house. The various bets are indicated in Figure 6-17 and are described in the following paragraphs:

Pass Line. You are betting *with* the dice, and the pay-off is even money. You win on a natural 7 or 11 on the first roll, you lose on "craps" 2, 3 or 12 on the first roll. Any other number on the first roll is the shooter's "point." You win if the "point" is thrown again, unless a 7 is thrown first, in which case you lose.

Don't Pass Line. Same as the *pass line*, except that you are betting

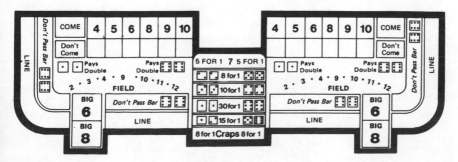

Figure 6-17. Craps Layout.

against the dice, and everything is reversed. You lose on a natural 7 or 11 on the first roll; you win on a "craps" 2, or 3. (When 12 is rolled it is a "stand-off"—nobody wins.) You lose after the first roll if the shooter makes his "point." You win if the shooter rolls 7 before making his "point."

Come. The simplest explanation of *come* bets is that you are betting *with* the dice exactly as on the pass line, except that *come* bets are made any time after the first roll. You win on "naturals" 7 or 11; you lose on "craps", 2, 3 and 12. Any number that comes up is the "come point." The bet is shifted to that number in the Craps layout. You win if your "come point" shows before a 7; otherwise you lose.

Don't Come. The play is again reversed. You are betting on the second roll but your bet is treated as if it were the first roll. You are betting *against* the dice exactly as on the *don't pass* line. You lose on naturals "7 or 11"; you win on "craps" 2 or 3. (When 12 is rolled, it's a "stand-off" . . . nobody wins). If a 4, 5, 6, 7, 8, 9 or 10 comes up on the second roll, you lose if the shooter makes that point on any subsequent rolls; you win if a 7 is rolled before the number shows again.

Field. You can bet on any roll that one of the following numbers comes up: 2, 3, 4, 9, 10, 11 or 12. If it does, you get even money, or two-to-one on 2 and three-to-one on 12, etc. If 5, 6, 7 or 8 comes up, you lose.

Big 6 or 8. You win even money if 6 or 8 shows before a 7 is rolled.

Any 7. You bet that the next roll is a 7 and you collect at four to one odds.

Any Craps. You bet that the next roll is 2, 3 or 12, and you collect at seven to one odds.

Hardways. You win if the exact combination you bet shows up. You lose if the same total number is rolled any other way except the hard way—or if a 7 comes up. Hardway combinations are:

DICE COMBINATION   PAYOFF

DICE COMBINATION	PAYOFF
⚀ ⚀	30 to 1
⚁ ⚁	7 to 1
⚂ ⚂	9 to 1

⚃ ⚃	9 to 1
⚄ ⚄	7 to 1
⚅ ⚅	30 to 1
⚀ ⚁	15 to 1
⚄ ⚅	15 to 1

A player may also place bets *with* or *against* a shooter's point. The payoff is made on whether or not the point shows before a seven.

Playing strategy consists of knowing the proper odds and not making a bet that is mathematically unsound. Computing the odds at Craps is simple, in most cases, since two dice can produce only 36 different combinations. These 36 dice combinations are illustrated in Figure 6-18, along with the probability of making each point. The odds for the principal bets are listed below and are determined by counting the number of combinations that will win for the player, and the number that will lose for the player.

Bet	Losing Chances	Winning Chances	Expectancy Of Loss	House Payoff Odds
PASS LINE	251	244	1.4%	Even
DON'T PASS	976	949	1.4%	Even
BIG 6 OR 8	6	5	9.1%	Even
FIELD	19	17	5.5%	Even
CRAPS (2,3 or 12)	32	4	11 %	7 to 1
SEVEN	30	6	16.7%	4 to 1
HARDWAY (2 or 12)	35	1	14 %	30 to 1
HARDWAY (4 or 10)	8	1	11 %	7 to 1
HARDWAY (6 or 8)	10	1	9 %	9 to 1
HARDWAY 11	34	2	11 %	15 to 1

In many cases the previous odds are different than those listed as house odds and winning dice combinations differ from house to house.

TOTAL	POSSIBLE COMBINATIONS	NUMBER OF WAYS	ODDS AGAINST
2	(dice combinations)	1	35–1
3	(dice combinations)	2	17–1
4	(dice combinations)	3	11–1
5	(dice combinations)	4	8–1
6	(dice combinations)	5	31–5
7	(dice combinations)	6	5–1
8	(dice combinations)	5	35–5
9	(dice combinations)	4	8–1
10	(dice combinations)	3	11–1
11	(dice combinations)	2	17–1
12	(dice combinations)	1	35–1

Figure 6-18. Thirty-six possible combinations of 2 dice.

6.11   Boule

La Boule (the bowl) is an extremely popular game in European casinos and Swiss resort hotels. La Boule is played with a stationary wheel that is divided into 18 compartments numbered from 1 to 9 twice. A *croupier* spins a small ball around the rim of the bowl. The ball will eventually come to rest in a number compartment which is considered to be the *winning number*. The number 5 is reserved for the house; therefore, the house expects to win 1/9th or 11+ percent of all money bet. Figure 6-19 shows a boule wheel that is used in European casinos.

Figure 6-20 illustrates a Boule layout. The following bets may be placed by players:

Bet	Payoff Odds
Any Number (1,2,3,4,5,6,7,8,9)	7 to 1
Odd Numbers, *impair* (1,3,7,9)	Pays even
Even Numbers, *pair* (2,4,6,8)	Pays even
First Four Numbers, *Manque* (1,2,3,4)	Pays even
Last Four Numbers, *Passe* (6,7,8,9)	Pays even
Black Numbers (1,3,6,8)	Pays even
Red Numbers (2,4,7,9)	Pays even

La Boule is popular, even in casinos where *Roulette* exists, because the minimum bet is low. Also the rules of Boule are very simple to learn.

Figure 6-19. Boule wheel.

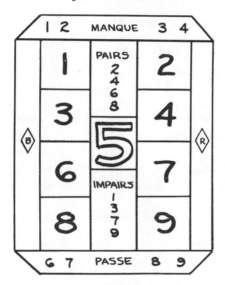

Figure 6-20. Boule layout.

6.12 Blackjack

Description of Blackjack. BLACKJACK* is the most popular card game in Nevada casinos. The game is often called *Twenty-one* (21) and is a game where everyone plays against a casino dealer. The dealer starts

---

* The French know it as *"Vingt-et-un"* (Twenty-one) and the English call it "Van John."

the game by dealing each player two cards, face down, and himself one card up and one card down. The object of the game is to draw cards that add up to 21 or as close to 21 as possible without going *bust*. You are *bust* when your cards total 22 or more.

Blackjack is played with either one standard 52-card deck or with four standard decks, which are dealt from a dealing device called a "shoe." The cards have no rank but have the following numerical value: Face cards and tens count 10, Aces count either 1 or 11, at the option of the player or dealer, and all other cards count at face value.

After the dealer has dealt the hand, you either *stand* (draw no more cards), or tell the dealer to *hit* (draw cards until you think the count of your hand is closer to 21 than the dealer will get). If your cards total over 21 you go *bust*. If your count is nearer 21 than the dealer's, you win. Ties are a *stand-off* and nobody wins.

If the first two cards of a player's hand count exactly 21, then the hand is called a *natural* or *Blackjack*, and the player will receive one and a half times his bet, unless the dealer also gets a Blackjack, in which case it is a stand-off and nobody wins. A Blackjack is any two-card Ace-10 combination. For example, the hand in Figure 6-21 is a Blackjack:

A player who wishes an additional card tells the dealer* to *HIT ME*. The dealer will then give the player one card face up. There is no limit to the number of cards** that a player may draw, except that he may not draw a card after he is *busted*. Figure 6-22 illustrates an original hand of 12 being hit with a face card and going *bust*:

If the dealer's original two cards total 16 or less, he must draw another card†. If his total is still less than 17, he must draw again until he either goes *bust* or totals 17 or greater. If the dealer's original two cards total 17 or more, he is not allowed to draw additional cards. For ex-

---

* A player may flick his cards to indicate to a dealer that he wishes his hand to be HIT.

** Gambling casinos have often offered special bonuses for player hands consisting of seven or more cards.

† Some casinos require the dealer to draw on a *soft 17*.

Figure 6-21. Blackjack hand.   Figure 6-22. Example of "busted" Blackjack hand.

Figure 6-23. Hands requiring a draw.

Figure 6-24. Hands not requiring a draw.

ample, the dealer would be required to draw on any one of the hands in Figure 6-23 and stand on any one of the hands in Figure 6-24.

If the dealer's hand goes over 21, he pays all bets not previously settled.

Figure 6-25 illustrates a typical layout that is printed on many Nevada blackjack tables.

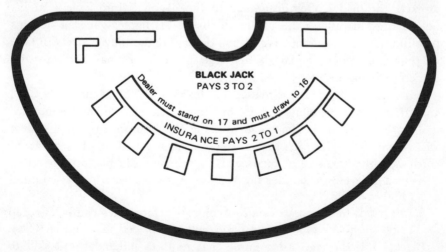

Figure 6-25. Blackjack layout.

Any two cards that are numerically identical (face cards have a value of 10) may be treated as a pair. The player has the option of *splitting pairs* and drawing to each hand. If the player elects to *split pairs*, then he must turn the cards face up and have the dealer hit each hand. Each hand may be played as normal hands—except in the following case. The player may only receive one card on each hand of Aces that were *split*.

Figure 6-26 shows two hands that may be split:

Figure 6-26. Hands that may be split.

When pairs are split, the player's original bet must be doubled to cover his twin hands.

The player may, after looking at his original cards, elect to double his bet and draw *one* additional card. This is known as *doubling down*. To *double down* the player must turn his original cards face up and place an equal amount of money beside the original bet. The dealer then gives the player one card, face down. The player's hand is represented by the total value of the three cards. If the dealer's up card is an Ace, the dealer offers *insurance* to the player before looking at his hole card. This allows the player to insure his hand against a dealer's possible natural. Insurance is not mandatory. If the player elects to take it, then an amount equal to half the player's original bet must be placed on the table.

When an insurance bet is made, the dealer will look at his down card, and, if it is a King, Queen, Jack or 10, turn it face up and announce that he has a natural. The player making the insurance bet will collect at a rate of 2 to 1 for each unit of his insurance bet. If the dealer's down card was other than 10 or face card, the player loses his insurance bet.

A *hard* hand either contains no Aces, or if it does, each Ace is counted as 1 (Figure 6-27).

A *soft* hand contains one or more Aces in which one Ace can be counted as 11 and the total count of the hand will not exceed 21 (Figure 6-28).

Blackjack is the one casino game that requires judgment and skill to play it well, however anyone can get the basic idea and play passably in a few minutes. Up until a few years ago, not even professional gamblers knew much about the game or its strategy. In recent years, however several mathematicians have, with the aid of computers, tested blackjack playing theories.

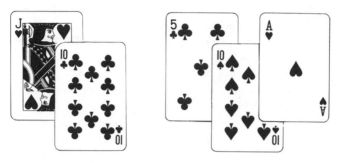

Figure 6-27. Examples of hard hands.

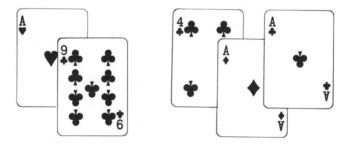

Figure 6-28. Examples of soft hands.

Professor Edward O. Thorp in his book *Beat The Dealer* made the gambling world take notice when he wrote of "how to beat the game of Blackjack." Thorp's method was based on card-counting and betting heavily when the cards remaining in the deck favored the player. *The Casino Gamblers Guide* by Dr. Allan Wilson, *The Theory of Gambling and Statistical Logic* by Richard A. Epstein, and *Playing Blackjack As A Business* by Lawrence Revere also contain blackjack strategies that were obtained with the use of computers.

Card counting is the most effective strategy for playing Blackjack. When the dealer begins a new hand without shuffling after the completion of the previous hand, the composition of the deck is different from the standard deck or decks he began with. This changes the probabilities of the hand totals which are possible on the next deal. By counting the discards after each hand, the player can exploit those situations in which he has a greater opportunity of winning by increasing the amount of his wager. He can also improve his playing strategy according to the remaining cards. Card counting is an effective playing strategy. It can reduce the casino advantage to a minimum.

Casino management is well aware of how some people play blackjack, and they don't like it. A person who counts the cards is known to the casino as *a counter*. Blackjack dealers are always on the alert to detect *counters*. If they suspect you of being a counter, they will attempt to distract your counting by speeding up the game, loudly riffling the cards, or by talking. If you are a consistent winner, they may even bar you from playing blackjack in their casino.

The basic strategy for playing Blackjack has been determined by several mathematicians. While these experts differ slightly on some points of the play, they are in surprisingly close agreement on the player's basic strategy. If the reader desires to learn more about the various blackjack strategies, it is suggested that he read one or more of the books by Thorp, Wilson, Epstein or Revere (see Bibliography).

In a single deck game in which the dealer must *hit* 16 and *stand* on 17, and the player is not allowed to *double down* after he has *split* a pair, the players strategy should usually be: Always *stand* on all hands with a count of 17 or more except a *soft* 17 and a *soft* 18. *Stand* on a count of 13 through 16 if dealer's up card is 2 through 6, but draw if dealer's card is 7 through 10, or an Ace. *Stand* on a count of 12 if the dealer's up card is a 4, 5, or 6. Otherwise, draw a card. Always *hit* a soft hand of 17 or less. *Hit* a soft 18 only if the dealer's face-up card has a count of 9 or 10.

Always *split* a pair of Aces or Eights. Never *split* a pair of face cards, tens, or fives. *Split* a pair of twos, threes, or sixes if dealer has 2 through 7 face up; with any other face-up card, *hit*. *Split* a pair of sevens if the dealer has 2 through 8 face up. *Split* a pair of nines if the dealer has 2 through 6, 8 or 9, face-up. *Split* fours if the dealer's up card is a 5.

Always *double down* with a count of 11. *Double down* with a count of 10 except when the dealer's face-up card is a 10 or Ace. *Double down* with 9 only when the dealer has a low face-up card, 2 to 6.

This over-all strategy can be used to play a good game of Blackjack.

Blackjack can be played with a small computer via a teletypewriter. In this type of system the computer program should be written in such a manner that it prints playing instructions to the player and directs all of the player's entries on the teletype. For example, the teletypewriter messages should be descriptive and should help the human player make his entries at appropriate times. The teletypewriter printout of Figure 6-29 illustrates the play of six games of Blackjack on a computer system. The computer wins two of the games and the player wins the remaining four.

```
THE RULES FOR THE GAME ARE AS FOLLOWS:
1. BLACKJACK PAYS DOUBLE.
2. PAIRS MAY NOT BE SPLIT.
3. DEALER HITS 16, STANDS ON 17.          —Instructions to Player
4. TIES ARE REPLAYED.
GOOD LUCK


*

YOU HAVE Q OF SPADES, 2 OF HEARTS.
DEALER SHOWS Q OF HEARTS
HIT?  YES 6 OF DIAMONDS                    —Game 1: Player Wins
HIT?  NO
DEALER HAS 7 OF HEARTS, Q OF HEARTS
DEALER'S SCORE IS 17. YOUR SCORE IS 18.
YOU WIN THIS ONE.              $1.00



YOU HAVE 8 OF SPADES, K OF CLUBS.
DEALER SHOWS A OF HEARTS
HIT?  NO                                   —Game 2: Computer Wins
DEALER HAS 10 OF HEARTS, A OF HEARTS
DEALER'S SCORE IS 21. YOUR SCORE IS 18.
DEALER TAKES THIS ONE.        $0.00


YOU HAVE A OF DIAMONDS, 10 OF DIAMONDS.
DEALER SHOWS 8 OF HEARTS
YOU HAVE A BLACKJACK          $2.00        —Game 3: Player Wins with Black-
                                                    jack

*

YOU HAVE J OF CLUBS, 8 OF SPADES.
DEALER SHOWS 3 OF HEARTS
HIT?  NO
DEALER HAS J OF HEARTS, 3 OF HEARTS
4 OF DIAMONDS                              —Game 4: Player Wins
DEALER'S SCORE IS 17. YOUR SCORE IS 18.
YOU WIN THIS ONE.             $3.00

YOU HAVE 10 OF SPADES, 5 OF DIAMONDS.
DEALER SHOWS 4 OF HEARTS
HIT?  YES 4 OF SPADES
HIT?  NO
DEALER HAS 3 OF SPADES, 4 OF HEARTS
J OF DIAMONDS                              —Game 5: Player Wins
DEALER'S SCORE IS 17. YOUR SCORE IS 19.
YOU WIN THIS ONE.             $4.00

YOU HAVE 5 OF CLUBS, 8 OF HEARTS.
DEALER SHOWS 2 OF HEARTS
HIT?  YES 2 OF SPADES
HIT?  YES A OF CLUBS                       —Game 6: Computer Wins
HIT?  NO
DEALER HAS 5 OF HEARTS, 2 OF HEARTS
5 OF SPADES
2 OF CLUBS
6 OF CLUBS
DEALER'S SCORE IS 20. YOUR SCORE IS 16.
DEALER TAKES THIS ONE.        $3.00
```

Figure 6-29. Printout of blackjack play on a computer.

CHAPTER 7

Board Games

7.1 Knight Interchange Game

DRAW A FLOWCHART and write a program that will determine the minimum number of moves that are necessary to make the white and black knights change places on the board shown in Figure 7-1. A knight's move consists of one square in a vertical or horizontal direction plus one square diagonally. The knight can also jump over any piece on the board. The interchange can be accomplished with 16 individual moves.

7.2 Dara

Dara is a board game played by the Dakarkari people of Nigeria, North Africa. The board is shown in Figure 7-2 and consists of 30 depressions made in the ground. Each player has 12 stones and places them one at a time in the holes in alternate turns of play. This play continues until both players have all of their stones on the board.

The players then alternately move a stone orthogonally to the next

Figure 7-1. Knight interchange board.

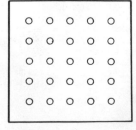

Figure 7-2. Dara board.

hole. The object of the game is to form three stones in a line in consecutive holes on the board.

Note that three diagonally arranged stones do not count. Whenever a player obtains three stones in a row or column, he may remove one of his opponent's stones from the board.

The game terminates when one player is unable to obtain a three-stone line.

7.3   The Knight's Tour

This problem is played with a standard 8 by 8 chessboard and a knight. The knight can move to a square that may be reached one square horizontally or vertically and one square diagonally away from the square occupied by the knight. Figure 7-3 illustrates the allowable moves that may be made by the knight. The square occupied by the knight is marked by a circle and the squares the knight may move to are marked by crosses.

The problem is to visit each square of the chessboard *once and only once* by using the knight's moves. This problem has drawn the attention of many prominent mathematicians for many years. However, the total number of solutions to this game has never been determined. Several Knight's Tours are shown in Figures 7-4, 7-5, and 7-6.

7.4   Five-field Keno

This game is played on the board shown in Figure 7.7. One player has 7 *black* stones and the other 7 *white* stones. The player with black stones always makes the first move. The players move one piece at a time, in alternate plays either backward or forward, or diagonally

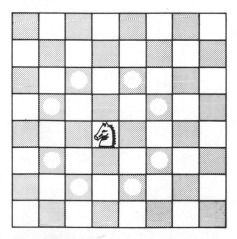

Figure 7-3. Moves of the knight.

47	10	23	64	49	2	59	6
22	63	48	9	60	5	50	3
11	46	61	24	1	52	7	58
62	21	12	45	8	57	4	51
19	36	25	40	13	44	53	30
26	39	20	33	56	29	14	43
35	18	37	28	41	16	31	54
38	27	34	17	32	55	42	15

50	11	24	63	14	37	26	35
23	62	51	12	25	34	15	38
10	49	64	21	40	13	36	27
61	22	9	52	33	28	39	16
48	7	60	1	20	41	54	29
59	4	45	8	53	32	17	42
6	47	2	57	44	19	30	55
3	58	5	46	31	56	43	18

Figure 7-4. Knight's Tours.

28	61	16	5	30	57	18	7
15	4	29	56	17	6	31	54
60	27	62	51	58	55	8	19
3	14	59	48	39	50	53	32
26	41	38	63	52	47	20	9
13	2	25	40	49	64	33	46
42	37	12	23	44	35	10	21
1	24	43	36	11	22	45	34

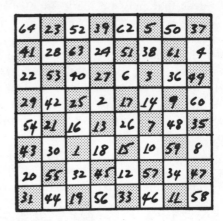

64	23	52	39	62	5	50	37
41	28	63	24	51	38	61	4
22	53	40	27	6	3	36	49
29	42	25	2	17	14	9	60
54	21	16	13	26	7	48	35
43	30	1	18	15	10	59	8
20	55	32	45	12	57	34	47
31	44	19	56	33	46	11	58

Figure 7-5. Knight's Tours.

1	42	13	26	3	60	15	28
24	37	2	41	14	27	4	61
43	12	25	38	59	62	29	16
36	23	40	63	48	51	56	5
11	44	49	52	39	58	17	30
22	35	64	47	50	55	6	57
45	10	33	20	53	8	31	18
34	21	46	9	32	19	54	7

Figure 7-6. A knight's tour.

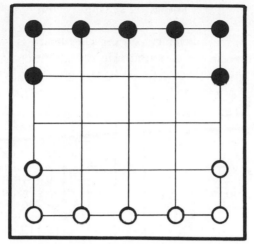

Figure 7-7. Board and starting position of stones in the game of five-field kono.

across the squares. The object of the game is to get the pieces across to the other side in the place of those pieces of the other player. The player who does this first wins the game.

7.5   The Queen's Journey

Figure 7-8 illustrates an 8 by 8 chessboard and the *queen* on her own square. The problem is to compute the greatest distance that the queen can travel over the chessboard in *five* queen's moves without passing over any square a second time. The queen must never cross her own track.

A correct solution to the problem is shown in Figure 7-9.

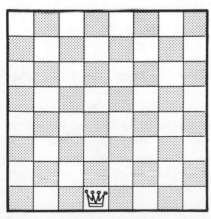

Figure 7-8. The queen in her starting position.

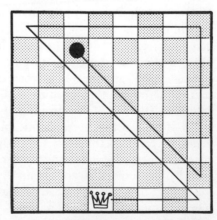

Figure 7-9. The queen's journey.

7.6   The King's Magic Tour

The *King's* power of movement on the chessboard is very limited. He can move only one square at a time. He can go into any of the squares —front, back or side—adjacent to the square on which he stands. Figure 7-10 illustrates the King's moves.

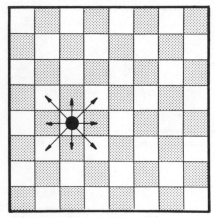

61	62	63	64	1	2	3	4
60	11	58	57	8	7	54	5
12	59	10	9	56	55	6	53
13	14	15	16	49	50	51	52
20	19	18	17	48	47	46	45
21	38	23	24	41	42	27	44
37	22	39	40	25	26	43	28
36	35	34	33	32	31	30	29

Figure 7-10. Moves of the King.   Figure 7-11. King's magic tour.

To complete a King's Tour one must move the King successively to every cell on the board.

Figure 7-11 illustrates such a tour. An interesting thing about this tour is that the numbers indicating the path form a *magic square*.

7.7   Go-Moko (Renjyu)

A study conducted at the Moore School of Electrical Engineering, University of Pennsylvania investigated the game of Go-Moko. The overall effort of this study was to attempt to broaden the areas in which computers can be useful in the solution of problems. One of the basic objectives of the study was to gain insight into methods by which computers might solve problems that cannot be explicitly stated in mathematical terms or that require deductive powers to give a satisfactory solution. The Go-Moko program was written by R. Wexelblat.

Go-Moko is a two-person game played on the 19 by 19 lined Go board. Each player has 180 stones; one player uses white stones, the other, black. Players alternate moves, placing a stone on an intersection of the board. The object is to obtain five adjacent stones in a row either vertically, horizontally, or diagonally. The player doing this wins the game. Figure 7-12 shows a game of Go-Moko where the player with white stones wins.

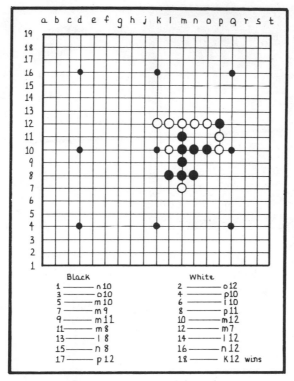

Figure 7-12. A game of Go-Moko.

7.8 Chess

Chess is a game in the play of which there is no element of chance. It is played on a board containing 64 squares, alternately colored black and white (or red and white). Each player has eight pieces and eight pawns. One series of pieces is light colored and called white, and the other is dark colored and called black. The men in each series consist of those shown in Figure 7-13.

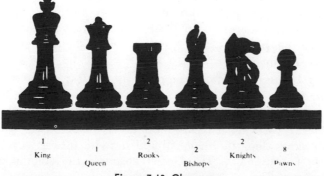

Figure 7-13. Chessmen.

Table 7-1. Chessmen Symbols.

	White King		Black King
♛	White King	♛	Black King
♛	" Queen	♛	" Queen
♝	" Bishop	♝	" Bishop
♞	" Knight	♞	" Knight
♜	" Rook	♜	" Rook
♟	" Pawn	♟	" Pawn

Figure 7-14. Starting positions of the chess pieces.

On commencing the game, the board should be set with a white square at the right-hand corner. The men shall be arranged on the chessboard as illustrated in Figure 7-14.

The symbols used on all chess diagrams represent the pieces shown in Table 7-1.

Those pieces on the Queen's side of the board are called the Queen's pieces, as the Queen's Rook, Bishop, etc., and those on the King's side, the King's pieces. The pawns likewise assume the name of the King or Queen together with the name of the pieces they stand in front of, as the King's Bishop's pawn, the King's pawn, etc.

The moves of the pieces are as follows:

1. *The King.* The King can move one square only at a time (except in *castling*), but he can make this move in any direction—forward, backward, laterally, or diagonally as shown in Figure 7-15a. He can take any one of the adversary's men which stands on an adjoining square to that he occupies, provided such man is left unprotected, and he has the peculiar privilege of being himself exempt from capture. He is not permitted, however, to move into check; that is, on to any square which is guarded by a piece or pawn of the enemy, nor can he, under any circumstance, be played to an adjacent square to that on which the rival King is stationed.

The King is the most important piece as the loss of the King means the loss of the game.

2. *The Queen.* The Queen is the most powerful of all the pieces. The Queen can move along the diagonal and straight, forward and sideways as shown in Figure 7-15b. The Queen possesses the combined powers of the Bishop and Rook.

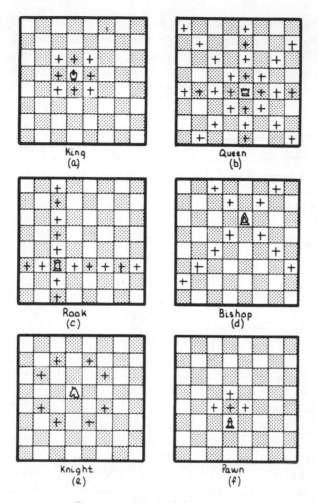

Figure 7-15. Moves of the chess men.

3. *The Rook.* The Rook is next in power to the Queen. He moves in a straight line, forward, backward, or sideways, having a uniform range, on a clear board, of 14 squares, exclusive of the one he occupies, as seen in Figure 7-15c.

The Rook has the same power in taking as the Queen—forward, backward and sideways, but he cannot, like her, take any man diagonally.

4. *The Bishop.* The Bishop moves diagonally forward or backward, to the extent of the board. The Bishop is next to the Rook in power. The two Bishops are originally placed on squares of different colors, and remain on these same colors throughout the game, as seen in Figure 7-15d.

5. *The Knight*. The Knight moves one square diagonally, then one forward, backward or sideways, or *vice versa*. He can move or capture in any direction, or can leap over his own men or any hostile man, and is the only piece that can play before any of the pawns have moved. Figure 7-15e describes the moves of the Knight. The Knight is next to the Bishop in power.

6. *The Pawn*. The Pawn moves forward only, and excepting its first move, only one square at a time. The Pawn can for its first move be moved two squares. The Pawn is the only man that does not capture as it moves forward. It captures on either of the two diagonal squares adjoining it in front, as seen in Figure 7-15f.

RULES AND MOVES.

1. The Chessboard must be so placed that each player has a white corner square nearest his right hand.
2. If a piece or pawn is misplaced at the beginning of the game, or the chessboard wrongly placed initially, the game shall be annulled.
3. The object of the game of Chess is to put the opposing King in such a position that he cannot ward off attack, either by interposing a piece or pawn or by moving to a place of safety. When this happens the game is over as the King is said to be *checkmated*.
4. The men must not be touched except by the correct player.
5. The first move shall be with the white men, and thereafter the players shall move alternately, one move at a time.
6. If a player can only move so as to take a pawn *en passant*, he must play that move.
7. The King is in check when he is attacked by a hostile piece or pawn; his capture is not permissible according to the rules of Chess.
8. *Stalemate* is brought about when the King although not at the moment in check, is so situated that he cannot be moved without going into check, and no other piece or pawn can be moved.
9. If neither the King nor the King's Rook has moved, and the squares between them are vacant, the King may move at once to his Knight's square and the Rook jump over the King and rest on the King's Bishop's square. This is called *castling* and is the one time in the game when the King may go two squares in a single move.
10. When any Pawn arrives at its eighth square, the player may substitute for it any other piece except the King.

Chess-playing computer programs can be simplified by using playing boards smaller than an 8 by 8 board. Figure 7-16 illustrates a 6 by 6 board that will allow all legal chess moves to be made. An even smaller board is shown in Figure 7-17.

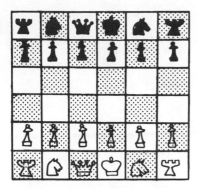

Figure 7-16. A 6 by 6 chessboard
configuration.

Figure 7-17. A 5 by 5 chessboard
and chess men.

7.9 Tic-Tac-Toe

Many machines have been designed and many computer programs
have been written to play Tic-Tac-Toe. One of the first machines was
conceived by Charles Babbage. Babbage, in his efforts to provide fund-
ing for the development of his Analytical Engine, conceived the idea of
machines that could play Chess and Tic-Tac-Toe. After further analyzing
the chess-playing machine, Babbage concluded that the possible combinations for a game of Chess surpassed the capabilities of his machine.
As soon as Babbage reached this conclusion he examined the game of
Tic-Tac-Toe. This game is much simpler than Chess and is played using
the following rules. One player uses an X marker and the second player
uses an O marker. The playing board consists of a square divided into
nine smaller squares, and the object of the player is to get three of his
own markers in a straight line. Plays are made on the board in alternate
moves. The moves of a typical game may be represented as shown in
Figure 7-18. In this example the player using marker X wins the game
in his fourth move.

Babbage had planned to build six machines so that they might be
exhibited in three different places at the same time. Note that even
Babbage believed in backup in case of machine failure. Babbage never
built these machines as he later felt that this enterprise would occupy

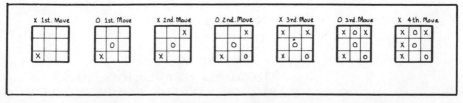

Figure 7-18. A game of Tic-Tac-Toe (player X wins on his 4th move).

so much of his time that he would not be able to use the obtained money to further develop the Analytical Engine.

The output of a typical Tic-Tac-Toe program might appear as follows:

HUMAN'S MOVE (1)
```
X  —  —
—  —  —
—  —  —
```

COMPUTER'S MOVE (2)
```
X  —  —
—  O  —
—  —  —
```

HUMAN'S MOVE (3)
```
X  —  —
—  O  —
—  —  X
```

COMPUTER'S MOVE (4)
```
X  O  —
—  O  —
—  —  X
```

HUMAN'S MOVE (5)
```
X  O  —
—  O  —
—  X  X
```

COMPUTER'S MOVE (6)
```
X  O  —
—  O  —
O  X  X
```

HUMAN'S MOVE (7)
```
X  O  —
X  O  —
O  X  X
```

COMPUTER'S MOVE (8)
```
X  O  O
X  O  —
O  X  X
```

COMPUTER WINS GAME
TRY AGAIN BY PRESSING THE START
BUTTON ON THE COMPUTER CON-
SOLE

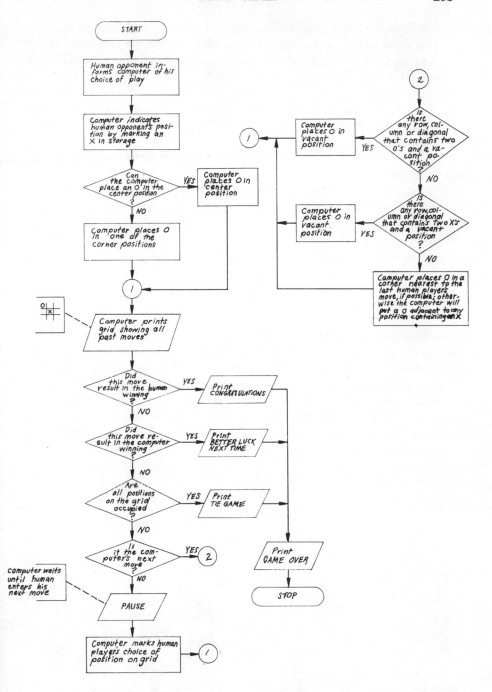

Figure 7-19. Flowchart of a Tic-Tac-Toe program.

A flowchart of the Tic-Tac-Toe Logic is shown in Figure 7-19.

This flowchart illustrates logic which allows either the computer or the human player to win. There is, however, a method of programming a computer where it is impossible for the human player to win. Can you modify the flowchart shown in Figure 7-19 to accomplish this task? It involves making two checks for special board situations after the program establishes that there are neither two X or O's in any row, column or diagonal. (Refer to Figure 12-19 for a flowchart of a Tic-Tac-Toe procedure that will allow the human player to tie but not to beat the computer).

Tic-Tac-Toe is a game where the programmer can become quite creative. There are many printer or typewriter formats that may be used and if the programmer is imaginative he can make the game output more interesting by producing appropriate messages at key game points. For example, the game printouts in Figure 7-20 illustrate an interesting format with comments.

7.10　Checkers

Checkers is played by two opponents with red and black counters, 12 for each opponent; the board is divided into 64 squares alternately colored black and white. The board should be placed between the opponents with white squares in the upper left-hand corner and double white squares in the lower left-hand corner. The counters of each opponent are placed on the first three rows of colored spaces, on opposite sides of the board, leaving two rows of colored spaces unoccupied between the counters of each player, as seen in Figure 7-21.

```
I'LL PLAY BLINDFOLDED                          I'M LEARNING

BOARD   YOUR PLAY    MY PLAY                    BOARD   YOUR PLAY    MY PLAY

1 2 3                                           1 2 3
4 5 6                                           4 5 6
7 8 9      5            3                       7 8 9      5            3

1 2 M                                           1 2 M
4 Y 6                                           4 Y 6
7 8 9      7            4                       7 8 9      7            4

1 2 M                                           1 2 M
M Y 6                                           M Y 6
Y 8 9      8            2                       Y 8 9      8            9

1 M M                                           1 2 M
M Y 6                                           M Y 6
Y Y 9      9                                    Y Y M      9 THAT'S MINE
                                                           2
1 M M                                           1 Y M
M Y 6                                           M Y 6
Y Y Y        GIVE THAT MAN A CIGAR             Y Y M     GIVE THAT MAN A CIGAR
```

Figure 7-20. Tic-Tac-Toe printouts.

```
I'M ON TO YOUR DIRTY TRICKS

BOARD   YOUR PLAY    MY PLAY

1 2 3
4 5 6
7 8 9        1            5

Y 2 3
4 M 6
7 8 9        9            3

Y 2 M
4 M 6
7 8 Y        7            8

Y 2 M
4 M 6
Y M Y        4

Y 2 M
Y M 6
Y M Y      FIRE THE PROGRAMMER
```

Figure 7-20. Tic-Tac-Toe printouts (cont'd).

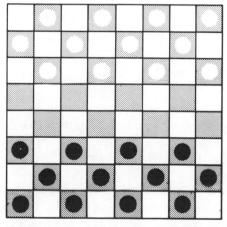

Figure 7-21. Checkerboard with starting positions of black and white men.

The counters are moved alternately on the dark squares only, in a diagonal forward direction, one square at a time. Whenever a counter progresses to the last row on the opposite side of the board, it is *crowned* by placing another counter on top of the counter that has reached this position. It is then called a *King*.

A King may move either backward or forward on the dark squares. The object of Checkers is to capture or block all of the opponent's men. Opponents make alternate moves and if an opponent can jump over the counter or counters of an opponent, one at a time, and find a resting-place on a dark square, the opponent loses the counters jumped in this way.

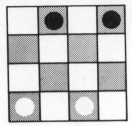

Figure 7-22. A 4 by 4 checkerboard with 4 men.

If a player does not capture an opponent's man whenever possible, the other opponent, if he detects the omission, may remove the opponent's counter, or he may compel his opponent to make the play.

The game is over whenever an opponent's counters are blocked so no moves can be made.

In order to simplify a check-playing program, one could use a much smaller board. The 4 by 4 board shown in Figure 7-22 may provide some interest to some readers. A program that will play checkers on this small board will contain much of the basic logic that would be used in an 8 by 8 board playing program.

7.11 Go

The ancient board game of Go was known in China thousands of years ago. It was introduced in Japan around 754 A.D. and has since become the most popular game in that country. Go is rapidly becoming more popular in the United States; however, it does not now have the popularity of Chess.

Go is a war game of skill between two opponents. It is played with small circular black and white *stones* on a 19 by 19 board. There are 361 intersection points on the grid. One player receives 181 black stones and the other receives 180 white stones. The black stones traditionally go to the weaker player. The player with black stones plays first, placing his stones on any intersection of the board, and thereafter the players alternately place one stone at a time on any vacant intersection point on the board which they desire to occupy. A stone once played cannot be moved; it remains where it is until it is captured or the game ends.

The object of the game of Go is to acquire *territory* by surrounding vacant points on the board with your stones.

In the course of play an opponent's stones may be surrounded and captured; however, that is not the primary object of the game. The player who has the highest score wins the game. A player's final score is determined by the number of vacant points that are surrounded by his stones and the number of captured stones in his possession.

The players alternately place stones on the board until the entire board is covered or until both players agree that one player has an advantage, and there is no point to continuing the game.

A stone is captured and removed from the board when all the points immediately adjacent to the stone have been occupied by enemy stones. Figure 7-23 illustrates the 19 by 19 Go board and several examples of white stones that have been captured and are ready for removal.

As shown in Figure 7-23, whenever a stone or stones are completely surrounded, both inside and outside by the opponent's stones, they are removed from the board. A stone cannot be placed on an intersection if the placing of the stone results in a capture of the stone.

Stones of like color are *connected* when they are adjacent to the same vertical or horizontal line. Two or more stones, so connected, are called a *unit*. A unit *lives* as long as it is connected to at least one vacant point. A unit *dies* when it can be completely enclosed.

A position such as *a* in Figure 7-24 is called Ko. If in such a situation the white stone moves into the position shown as *b* in Figure 7-24, he captures the black stone. If then the black stone were replaced in the previous position, he would capture a white stone and the pattern would simply oscillate. The rule of Ko prevents a player who has lost a single stone to capture, in the subsequent move, just the one stone by which it was taken. This one-play delay gives the opponent a chance of stopping the Ko.

A player may prefer to pass on his turn and this is permissible. He may not, however, pass on two consecutive turns of play.

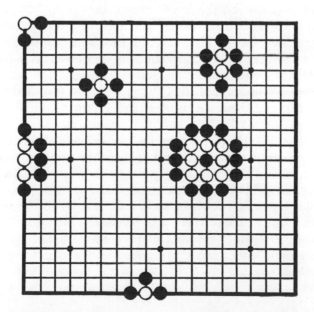

Figure 7-23. The Go board with groups of captured white stones.

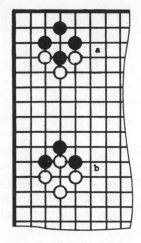

Figure 7-24. A Ko situation.

Figure 7-25. The mill board.

A 19 by 19 Go board permits approximately 10^{172} different board positions in the course of a game. In order to simplify the game for computer solution, one could use a smaller board. A 13 by 13 or 11 by 11 board is adequate for this purpose. There are approximately 10^{57} board positions on the 11 by 11 board.

7.12   Mill

The Mill is played on the board shown in Figure 7-25. It is a two player game where each player has nine men of a distinguishable color. The players draw for the first move, and each in turn places a man on any one of the corners or intersections of the lines. As soon as either player gets three men in line, he can remove from the board any one of his opponent's pieces, provided the piece is not one of three that are already in a line. If there are no other men on the board, he can remove one of a line.

After placing all nine of his men, the player can shift them about, one at a time, but no man can be moved farther than to the adjoining corner or intersection, and that must be vacant. As soon as a new line is formed in this way, an opponent's man can be removed. It is possible for one man to move back and forth in such a manner as to continually form and reform a line of three. When one player has only three men left, he can jump any of them to any vacant space on the board, no matter how far off. As soon as either player is reduced to two men, the game is over and he has lost.

CHAPTER 8

Magic Squares

8.1 Card Magic Squares

A 3 BY 3 MAGIC SQUARE constructed of playing cards is shown in Figure 8-1.

The relationship between this 3 by 3 card square and a regular 3 by 3 Magic Square is a one-to-one substitution between a value of the regular magic square and the corresponding card value of the card square. The card merely represents the corresponding number values of a regular magic square.

Card magic squares can be generated using any of the methods of Odd and Even magic square generation noted in Chapter 2.

The following card values are used in generating the card squares:

Card Value	Magic Square Value
10,J,Q,K	10
9	9
8	8
•	•
•	•
•	•
2	2
A	1

Figure 8-2 illustrates a 4 by 4 card square and a 5 by 5 card square is shown in Figure 8-3.

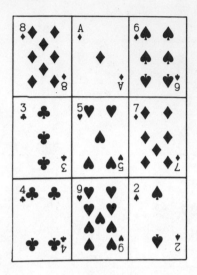

Figure 8-1. A 3 by 3 card magic square.

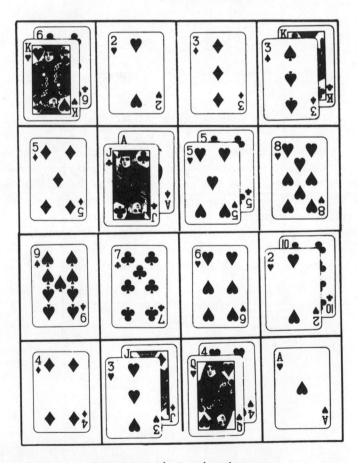

Figure 8-2. A 4 by 4 card magic square.

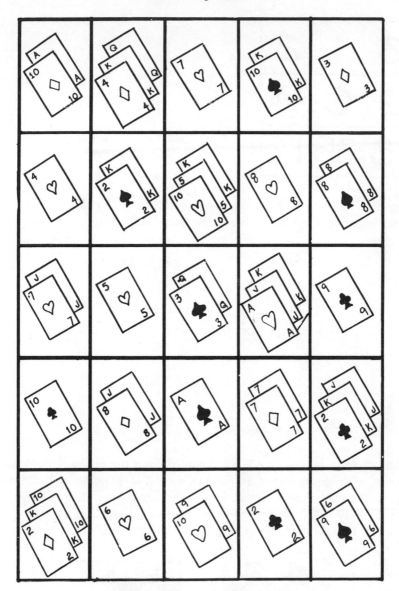

Figure 8-3. A 5 by 5 card magic square.

8.2 Division Magic Square

A 3 by 3 Division Magic Square is illustrated in Figure 8-4.

The magic number of this square is 6 and is obtained by dividing the second number in any row, column or diagonal by the first number in the respective row, column or diagonal. The quotient of this division is then divided into the third number of the respective row, column or

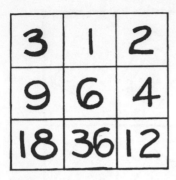

Figure 8-4. A 3 by 3 division magic
square.

diagonal. This division process is simplified by dividing the *product* of the extreme numbers by the center number. Thus

$$ROW\ 1 \qquad = (3 \times 2)/1 = 6$$

$$ROW\ 2 \qquad = (9 \times 4)/6 = 6$$

$$ROW\ 3 \qquad = (18 \times 12)/36 = 6$$

$$COLUMN\ 1 \quad = (3 \times 18)/9 = 6$$

$$COLUMN\ 2 \quad = (1 \times 36)/6 = 6$$

$$COLUMN\ 3 \quad = (2 \times 12)/4 = 6$$

$$DIAGONAL\ 1 = (3 \times 12)/6 = 6$$

$$DIAGONAL\ 2 = (2 \times 18)/6 = 6$$

The division square may be constructed from the *multiplication magic square* by reversing the main diagonals, and exchanging opposite numbers in the center cells of each border row or column. For example, the 5 by 5 Division Magic Square of Figure 8-5a may be derived from the 5 by 5 Multiplication Magic Square of Figure 8-5b by changing the

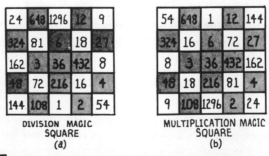

DIVISION MAGIC
SQUARE
(a)

MULTIPLICATION MAGIC
SQUARE
(b)

cell contents remain the same in both squares

Figure 8-5. Construction of a 5 by 5 division magic square.

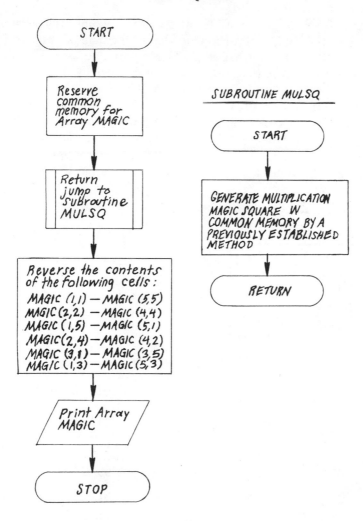

Figure 8-6. Flowchart of the division magic square program.

contents of 12 cells. Using subscript notation and assuming that a Multiplication Magic Square was contained in Array MAGIC, the cells needed to be switched are listed as follows:

DIAGONAL 1 { MAGIC (1,1) ...... MAGIC (5,5)
MAGIC (2,2) ...... MAGIC (4,4)
DIAGONAL 2 { MAGIC (1,5) ...... MAGIC (5,1)
MAGIC (2,4) ...... MAGIC (4,2)
ROW 3   MAGIC (3,1) ...... MAGIC (3,5)
COLUMN 3   MAGIC (1,3) ...... MAGIC (5,3)

A FORTRAN program may be written to generate a Division Magic Square by making a subroutine of the *Multiplication Magic Square Generating Program* of Chapter 2. This subroutine would be used to generate a 5 by 5 Multiplication Square and it would only be necessary to reverse the specified cell locations. A flowchart of this program is shown in Figure 8-6.

Note that is is necessary to place Array MAGIC in COMMON memory, and the subroutine should not PRINT the Multiplication Magic Square.

Figure 8-7 illustrates a 4 by 4 Division Magic Square. This square has a magic number of 9.

27	2	72	108
54	1	4	24
8	12	3	18
36	216	6	9

Figure 8-7. A 4 by 4 division magic square.

8.3 Upside-Down Magic Square

The magic number of the following 4 by 4 array of numbers is 264:

96	11	89	68
88	69	91	16
61	86	18	99
19	98	66	81

In addition to the rows, columns and diagonals having a sum of 264, there are additional number combinations that add up to the same value. If each number of the above magic square was given a corresponding location in the following square, then the number combinations in Figure 8-8 would have a sum of 264.

A different number arrangement is obtained by turning the magic square upside down. This is illustrated in the following square:

18	99	86	61
66	81	98	19
91	16	69	88
89	68	11	96

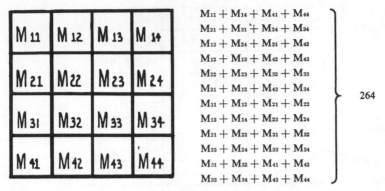

$$M_{11} + M_{14} + M_{41} + M_{44}$$
$$M_{21} + M_{31} + M_{24} + M_{34}$$
$$M_{13} + M_{24} + M_{31} + M_{42}$$
$$M_{12} + M_{13} + M_{42} + M_{43}$$
$$M_{22} + M_{23} + M_{32} + M_{33}$$
$$M_{21} + M_{12} + M_{43} + M_{34}$$
$$M_{11} + M_{12} + M_{21} + M_{22}$$
$$M_{13} + M_{14} + M_{23} + M_{24}$$
$$M_{21} + M_{22} + M_{31} + M_{32}$$
$$M_{23} + M_{24} + M_{33} + M_{34}$$
$$M_{31} + M_{32} + M_{41} + M_{42}$$
$$M_{33} + M_{34} + M_{43} + M_{44}$$

$$\left.\begin{array}{c}\end{array}\right\} 264$$

Figure 8-8. Upside-down magic square.

If the reader investigates the above upside-down square, he will find that each row, column, diagonal and position listed for the previous magic square are all equal to 264.

8.4 Composite Magic Square

Another type of magic square is the *composite*, an example of which is shown in Figure 8-9 where a series of small magic squares is arranged in magic square order. This 9 by 9 square is the smallest composite square and is composed of nine 3 by 3 magic squares arranged in the order where a would be replaced by a 3 by 3 square starting with the number 1; b would be replaced by a 3 by 3 square starting with 10; c would be replaced by a square starting with 19, and so forth, until i would be replaced by a square starting with 73.

An *Order 12 Composite Square* may be constructed of sub-squares of even order (Order 4). The 4 by 4 squares would be placed in the same order as were the 3 by 3 squares in the Order 9 Composite Square. The 12 by 12 Composite Square is shown in Figure 8-10.

h	a	f
c	e	g
d	i	b

71	64	69	8	1	6	53	46	51
66	68	70	3	5	7	48	50	52
67	72	65	4	9	2	49	54	47
26	19	24	44	37	42	62	55	60
21	23	25	39	41	43	57	59	61
22	27	20	40	45	38	58	63	56
35	28	33	80	73	78	17	10	15
30	32	34	75	77	79	12	14	16
31	36	29	76	81	74	13	18	11

Figure 8-9. A 9 by 9 composite magic square.

113	127	126	116	1	15	14	4	81	95	94	84
124	118	119	121	12	6	7	9	92	86	87	89
120	122	123	117	8	10	11	5	88	90	91	85
125	115	114	128	13	3	2	16	93	83	82	96
33	47	46	36	65	79	78	68	97	111	110	100
44	38	39	41	76	70	71	73	108	102	103	105
40	42	43	37	72	74	75	69	104	106	107	101
45	35	34	48	77	67	66	80	109	99	98	112
49	63	62	52	129	143	142	132	17	31	30	20
60	54	55	57	140	134	135	137	28	22	23	25
56	58	59	53	136	138	139	133	24	26	27	21
61	51	50	64	141	131	130	144	29	19	18	32

Figure 8-10. A 12 by 12 composite magic square.

A *15 by 15 Composite Magic Square* may be constructed using nine 5 by 5 squares. An *Order 18 Composite Square* may be constructed by combining nine 6 by 6 magic squares in the specified order.

8.5   Odd Order Magic Square

The *Bachet de Meziriac* method for generating odd order magic squares is similar to the *De la Loubere* method discussed in Chapter 2. The two differences are:

(1) The starting point is located over the center of the center square instead of in the top row.

(2) When the end of a group of n numbers is reached, the new position is in the same column but 2 cells higher.

1. Place the number in the square just above the center square.
2. Place succeeding numbers on the diagonal leading to the right and up. Follow this rule whenever possible.
3. If you leave the square at the top, go to the bottom of the column in which you want to place the number.
4. If you leave the square at the right, go to the extreme left of the row in which you wanted to place the number.
5. If the space in which you wish to place a number already contains a number, go back to the square you left and put the number in the square two spaces above that square. If you leave the top of the

square, visualize the top and bottom squares as being attached to each other.

6. When you leave the square out of the main diagonal going from bottom left to top right, place the next number in the extreme right of the next to last row, and then follow rule 4, etc.

7. The magic square will be complete when the last number is placed in the location just below the center cell.

A 5 by 5 Magic Square is shown in Figure 8-11. This square was generated by the *Bachet de Meziriac* method.

23	6	19	2	15
10	18	1	14	22
17	5	13	21	9
4	12	25	8	16
11	24	7	20	3

Figure 8-11. A 5 by 5 magic square.

Table 8-1 lists the numbers of the 5 by 5 Magic Square of Figure 8-11 and the associated rules needed for placing the numbers.

Other magic squares generated by the *Bachet de Meziriac* method are shown in Figure 8-12.

Table 8-1. Rule numbers for Generating a 5 by 5 Magic Square.

Number	Rule	Number	Rule
1	1	13	2
2	2	14	2
3	3	15	2
4	4	16	6
5	2	17	4
6	5	18	2
7	3	19	2
8	2	20	3
9	2	21	5
10	4	22	2
11	5	23	4
12	2	24	3
		25	2

8	1	6
3	5	7
4	9	2

46	15	40	9	34	3	28
21	39	8	33	2	27	45
38	14	32	1	26	44	20
13	31	7	25	43	19	37
30	6	24	49	18	36	12
5	23	48	17	42	11	29
22	47	16	41	10	35	4

77	28	69	20	61	12	53	4	45
36	68	19	60	11	52	3	44	76
67	27	59	10	51	2	43	75	35
26	58	18	50	1	42	74	34	66
57	17	49	9	41	73	33	65	25
16	48	8	40	81	32	64	24	56
47	7	39	80	31	72	23	55	15
6	38	79	30	71	22	63	14	46
37	78	29	70	21	62	13	54	5

Figure 8-12. Odd order magic squares generated by the Bachet de Meziriac method.

The solution of a *Bachet de Meziriac Magic Square Generating Program* is similar to the *De la Loubere Magic Square Generating Program* which was discussed in Chapter 2.

8.6 Multiplication Semi-Magic Square

A *Multiplication Semi-magic* Square* is an arrangement of numbers in a square so that all rows and columns have an equal product. A 3 by 3 Multiplication Semi-magic Square is shown in Figure 8-13.

Each row and column of this square has a product of 120.

ROWS $8 \times 15 \times 1 = 5 \times 2 \times 12 = 3 \times 4 \times 10 = 120$
COLUMNS $8 \times 5 \times 3 = 15 \times 2 \times 4 = 1 \times 12 \times 10 = 120$

The product of the diagonals does not equal 120.

DIAGONAL 1: $8 \times 2 \times 10 = 160$
DIAGONAL 2: $1 \times 2 \times 3 = 6$

8.7 Diamond Magic Square

Another popular method for forming odd order magic squares is to arrange the square from a set of numbers arranged in a natural square. Suppose the first 25 numbers are to be arranged in a magic square. First place the numbers in the form of a *Natural Square,* as shown in

---

* A square is semi-magic if one or both of the main diagonal sums differ from the magic number.

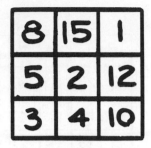

Figure 8-13. Multiplication semi-magic square.

Figure 8-14. Next draw 4 straight lines, cutting off three numbers at each corner: 1, 2 and 6 from the left hand upper corner; 16, 21 and 22 from the lower left hand corner; 4, 5 and 10 from the right hand upper corner; and 20, 24 and 25 from the lower right hand corner.

These four lines form a large square. Divide this large square into 25 cells by drawing inner lines parallel to the sides of the large square. Thirteen of these cells will be occupied by numbers, as shown in Figure 8-15.

The empty cells in Figure 8-15 are to be filled using the corner numbers that were left outside the large square: each number from the corners is to be moved obliquely up or down along the row or column where it is found, to the remote vacant cell of the large square. The magic square in Figure 8-16 is how the square would appear after filling in the vacant squares.

8.8   De La Hire Magic Square

De la Hire devised a method that may be used for constructing magic squares of any order. Figure 8-17 illustrates a square that was constructed by this method.

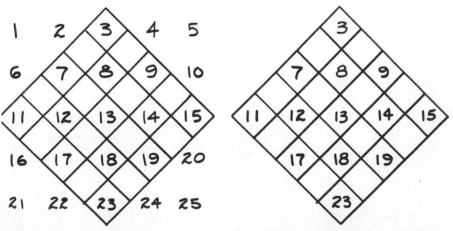

Figure 8-14. Twenty-five numbers in natural order.

Figure 8-15. Part of a 5 by 5 magic square.

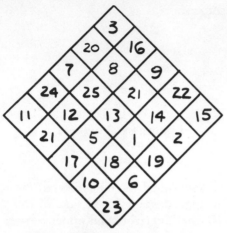

Figure 8-16. An order 5 magic
square formed by the diamond
generating method.

18	1	24	7	15
5	23	6	14	17
22	10	13	16	4
9	12	20	3	21
11	19	2	25	8

Figure 8-17. A 5 by 5 magic square
constructed using the *De la Hire*
method.

De la Hire's method uses three squares in the generation of a magic
square. One of these squares contains the numbers 1, 2, 3, 4 and 5
arranged so that every number appears once and only once in each row,
column and main diagonal 2. Main diagonal 1 contains the number 3
in each cell located on the diagonal line. The second square contains
the numbers 0, 5, 10, 15 and 20 and they are arranged in a manner
similar to the first square except 10's are placed in the cells lying on
diagonal 2. The third square is a blank square. The three squares are
illustrated in Figure 8-18.

A magic square is produced in the *blank square* by placing in each
cell the sum of the corresponding cells of *Square 1* and *Square 2*. The
reader may construct the square shown in Figure 8-17 by performing
the required addition.

Figure 8-19 illustrates a flowchart that may be used for generating
the *De la Hire* magic square.

8.9   Prime Magic Squares

The 3 by 3 Magic Square of Figure 8-20 is constructed of *prime numbers*. The square was first constructed by the British puzzle expert Henry

3	1	4	2	5
5	3	1	4	2
2	5	3	1	4
4	2	5	3	1
1	4	2	5	3

Square 1

15	0	20	5	10
0	20	5	10	15
20	5	10	15	0
5	10	15	0	20
10	15	0	20	5

Square 2

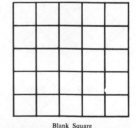

Blank Square

Figure 8-18. Three squares used in the *De la Hire* magic square construction method.

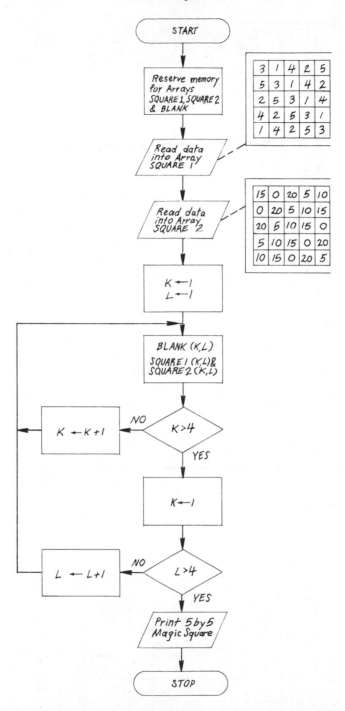

Figure 8-19. Flowchart for a De la Hire magic square generating program.

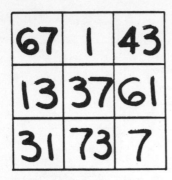

Figure 8-20. A 3 by 3 prime magic square.

Ernest Dudeney. The sum of each row, column and diagonal is 111.

The reader should note that the numbers of this square are not consecutive prime numbers. That is, several prime numbers in the range 1 to 73 have been omitted: 2, 3, 5, 11, 17, 19, 23, 29, 41, 47, 53, 59, 71.

A 12 by 12 Prime Magic Square consisting of the first 144 consecutive odd prime numbers was generated by J. N. Muncey in 1913. This square is the smallest magic square that can be generated from consecutive prime numbers. The one *even* prime number, 2, cannot be considered in generating a prime magic square because it would change the polarity of the row, column or diagonal that it appeared in. Figure 8-21 illustrates the 12 by 12 Prime Magic Square. This square has a *magic number* of 4514.

Prime magic squares have been constructed for all orders up to and

827	659	661	509	229	503	349	367	223	97	89	1
3	673	101	199	491	523	359	379	653	227	83	823
7	677	643	73	373	233	353	521	499	103	211	821
5	683	239	541	487	337	647	383	197	107	79	809
13	71	691	347	461	547	389	241	109	193	641	811
11	67	701	191	251	397	331	467	113	557	631	797
787	61	127	181	443	421	317	257	563	719	619	19
769	47	131	569	463	17	311	263	479	727	709	29
773	59	179	577	137	401	409	269	173	607	617	313
419	743	613	571	439	271	307	167	761	139	53	31
149	733	277	163	457	431	293	601	587	757	43	23
751	41	151	593	283	433	449	599	157	281	739	37

Figure 8-21. A 12 by 12 prime magic square.

including Order 12. The magic numbers of the first ten prime magic squares are:

Order of Square	Magic Number
3	111
4	102
5	213
6	408
7	699
8	1114
9	1681
10	2416
11	3355
12	4514

Write a program to prove that the 12 by 12 square of Figure 8-21 is indeed a magic square.

8.10 Subtracting Magic Squares

A *3 by 3 Subtracting Magic Square* is shown in Figure 8-22. This square has a magic number of 5 and is determined by subtracting the first number of any row, column or diagonal from the second number, and the result from the third number of the same row, column or diagonal. The magic number of the odd order subtracting magic square may also be calculated by the formula

$$\frac{N(N^2+1)}{2N}$$

where N is the order of the square.

Figure 8-23 illustrates a 5 by 5 Subtracting Magic Square with a magic number of 13.

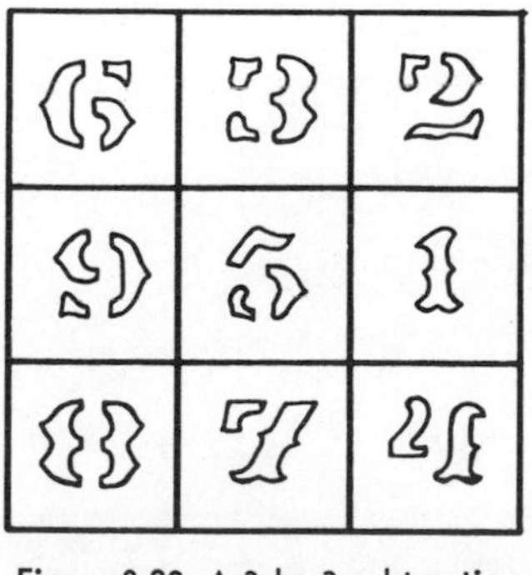

Figure 8-22. A 3 by 3 subtracting magic square.

9	24	25	8	11
23	21	7	12	16
22	6	13	20	4
10	14	19	5	3
15	18	1	2	17

Figure 8-23. A 5 by 5 subtracting magic square.

$$\frac{\text{MAGIC}}{\text{NUMBER}} = \frac{N(N^2+1)}{2N} = \frac{5(5^2+1)}{2 \times 5} = 13$$

This square was derived from a regular 5 by 5 Magic Square generated by the *De la Loubere* method. It is necessary to reverse the numbers lying in cells on the main diagonals and exchange the numbers in opposite center border cells. Using subscript notation and assuming that a *De la Loubere* magic square was contained in Array KMAG, the cells needed to be exchanged are listed below.

KMAG(1,1) ...... KMAG(5,5)
KMAG(2,2) ...... KMAG(4,4)
KMAG(1,5) ...... KMAG(5,1)
KMAG(2,4) ...... KMAG(4,2)
KMAG(1,3) ...... KMAG(5,3)
KMAG(3,1) ...... KMAG(3,5)

A 4 by 4 Subtracting Magic Square is shown in Figure 8-24. The magic number of this square is 8.

9	6	16	11
10	5	7	4
8	3	1	14
15	12	2	13

Figure 8-24. A 4 by 4 subtracting magic square.

CHAPTER 9

Number Games

9.1   Perfect Number

THE NUMBER 6 HAS a curious property. By adding the divisors of 6, a sum equal to the number itself is found:

$$3+2+1 = 6$$

This is also true of the number 28.

$$14+7+4+2+1 = 28$$

These numbers are called *perfect*. An *odd* perfect number has never been found, but no proof exists that says that an odd number cannot be perfect. Euclid's formula for generating perfect numbers is

$$2^{n-1}(2^n-1)$$

in which the factor 2^n-1 must be a *prime number*. For many years only 12 perfect numbers were known, namely those computed by setting n equal to 2, 3, 5, 7, 13, 17, 19, 31, 61, 89, 107 and 127 in Euclid's formula.

n	$2^{n-1}(2^n-1)$
2	6
3	28
5	496
7	8128
13	33,550,336
17	8,589,869,056
.	.
.	.
.	.

In recent years the digital computer has computed several additional perfect numbers. One large one is

$$2^{2280}(2^{2281}-1)$$

9.2 Date—Day Game

The following rules are used to play the date—day game:

1. Pick a date.
2. Add ¼ of the last two digits of the year to the last two digits of the year.
3. Add the month number shown in Table 9-1b.
4. Add the day.
5. Add the year number shown in Table 9-1a.
6. Divide by 7.
7. The day is found in Table 9-1c. by using the remainder.

Example:

What day did September 3, 1848, fall on?

1. The date is September 3, 1848.

Table 9-1. Date-day Game Table.

(a) Year Table

1900 — 2000	Add 0
1800 — 1900	Add 2
9/14/1752 — 1800	Add 4
1700 — 9/2/1752	Add 1
1600 — 1700	Add 2

(b) Month Table

January	Add 1	July	Add 0	Leap Year	0
February	Add 4	August	Add 3	Leap Year	3
March	Add 4	September	Add 6		
April	Add 0	October	Add 1		
May	Add 2	November	Add 4		
June	Add 5	December	Add 6		

(c) Day Table

Sunday	1	Wednesday	4
Monday	2	Thursday	5
Tuesday	3	Friday	6
	Saturday	0	

2. $\tfrac{1}{4}(48) = 12$ $48+12 = 60$
3. $60+6 = 66$
4. $66+3 = 69$
5. $69+2 = 71$
6. $71 \div 7 = $ Quotient $= 10$ Remainder $= 1$
7. Remainder 1 specifies the day *Sunday*.

9.3   Figurate Numbers

Figure 9-1 illustrates *Pascal's triangle*, which is an orderly way of writing the *figurate numbers*. The numbers are called *figurate* because early mathematicians thought that they gave the areas and volumes of certain geometrical figures when built up by discrete units.

The top row and left most column of Pascal's triangle consists entirely of 1's. The rest of the table is made up by writing in each square the sum of the two numbers in the square at its left and above it.

The reader may recognize that the numbers that lie along a diagonal line are the coefficients of the terms in a *binomial* expansion. For example:

$$(a+b)^2 \qquad 1a^2+2ab+1b^2 = \qquad\qquad\qquad 1,2,1$$
$$(a+b)^3 \qquad 1a^3+3a^2b+3b^2a+1b^3 = \qquad\qquad 1,3,3,1$$
$$(a+b)^4 \qquad 1a^4+4a^3b+6a^2b^2+4ab^3+1b^4 = 1,4,6,4,1$$

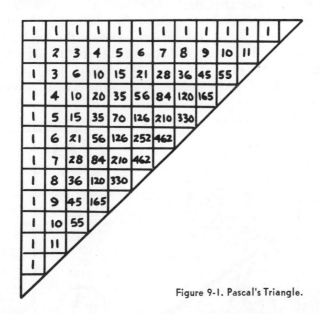

Figure 9-1. Pascal's Triangle.

CHAPTER 10

Unusual Gambling Games

10.1 Chinese Fan Tan

THE BANKING GAME of *Chinese Fan Tan* is popular throughout the East, especially in China and Korea. The game is played by a banker placing an unknown number of beans, coins, dice, poker chips or other small objects under a Chinese soup bowl. Figure 10-1 illustrates a Fan Tan layout along with several beans, a bowl and a counting stick. Players may place their bets on a single number, or between any two numbers.

After all bets are placed, the banker removes the bowl and counts off the beans, four at a time, with the counting stick. The number of beans left decides the bets.

Bets on a single number pay 3 for 1. Bets between two numbers and

Figure 10-1. Chinese fan tan layout with beans, bowl and counting stick.

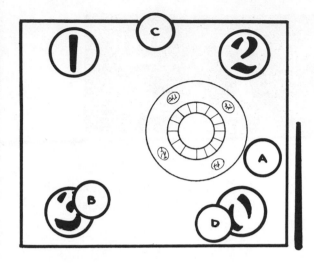

Figure 10-2. Player bets in Chinese fan tan.

on double numbers pay even money. For example, assume players A, B, C and D placed bets on the Fan Tan layout as shown in Figure 10-2.

After removing the bowl and reducing the bean pile, modulo 4, as shown in Figure 10-3, only three beans remain. Bets would be paid off as follows:

Player A, who bet on 2 and 0, would lose his bet.
Player B, who bet on 3, would be paid at 3 to 1 odds.
Player C, who bet on 1 and 2, would lose his bet.
Player D, who bet on 0, would lose his bet.

10.2 Under and Over Seven

The *Under and Over Seven*, sometimes called *Over and Under*, is found in small gambling joints rather than plush gambling casinos. In

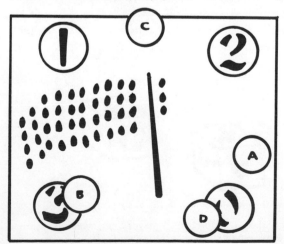

Figure 10-3. Counting the beans in Chinese fan tan.

Figure 10-4. Over and under seven layout.

this game, two dice are thrown, and players bet that the total will be *Over* 7, *Under* 7 or *Equal* to 7. Figure 10-4 illustrates the layout used in the *Under and Over Seven* game.

A player places his bet on the layout, and after the roll of the dice, is either paid off at 3 to 1, even money or his bet is lost. A bet on *Over* 7 wins when the roll of the dice produced a sum of 8, 9, 10, 11 or 12. An *Under* 7 bet would be a winner if the dice total was 2, 3, 4, 5 or 6. Both the *Over* 7 and *Under* 7 bets are paid off at even money. If the player puts a bet in the center of the layout and the dice total was 7, he would be paid off at odds of 3 to 1.°

The bank's advantage is a large one since the odds are 5 to 1 against 7 and 21 to 15 against *Under* 7 or *Over* 7.

10.3   Crown and Anchor

The *Crown and Anchor* game uses three dice. Each dice has four card symbols plus a crown and an anchor on its six faces. (See Figure 10-5.)

Figure 10-6 illustrates the Crown and Anchor layout. Each symbol on the layout is represented on the dice.

A player may bet on any one of the six symbols and will be paid even money if the selected symbol shows once, double if it shows twice, and triple if it shows on all three dice.

Figure 10-5. Dice for crown and anchor game.

° 3 to 1 is written on some layouts as 4 for 1. The player's bet, which is returned to him, is counted in the latter payoff.

Figure 10-6. Crown and anchor layout.

10.4 Beat the Dealer

Beat the Dealer is a banking game played in small gambling houses. The game uses a pair of dice, a dice cup and a layout similar to the one illustrated in Figure 10-7.

The game is started by the banker throwing the dice from a dice cup. The value the banker throws is marked on the layout with a poker chip, coin, checker or some other small object. Now it is the player's turn to throw the dice. The player can win the game only if he throws a higher value than that marked on the layout. The banker wins on all ties.

10.5 Poker Dice

The *Poker Dice* game uses five dice, each marked from ace to nine (Figure 10-8).

Figure 10-7. Beat the dealer layout.

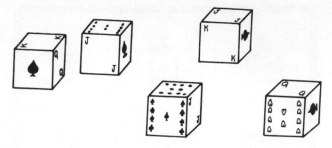

Figure 10-8. Poker dice.

The object of the game is to roll the dice and to make the best Poker hand, in one, two, or three rolls. Any number of players may play the game. The rank of hands is:

1. Five of a kind
2. Flush
3. Four of a kind
4. Straight
5. No pair or no sequence
6. Five cards in sequence (Q, J, 10, 9, A)
7. Three of a kind
8. Two pair
9. One pair

The first player tosses the five dice. He may accept the toss, or put any of the dice aside and toss the others again. After his second toss, he may put aside any of the dice, adding them to the ones previously put aside, and toss the remaining die or dice a third time. The first player's hand is now recorded and the next player tosses the dice in a similar manner.

Betting usually consists of each player putting a chip in a common pot prior to tossing the dice. The player with the highest hand wins the pot.

CHAPTER 11

Miscellaneous Games

11.1 Nim

THE WELL KNOWN GAME of *Nim* is an ancient two-person mathematical game. This game, based on binary principles, could be played somewhat easier if one possessed a binary adder.

Nim is played by two persons, using matches, coins, sticks, beads, poker chips, or any other objects as counters. The counters are put into piles and the players take turns removing the counters from the piles. The number of counters in each pile, and the number of piles can vary with the game.

Each player in his turn must remove at least one counter from a pile. More than one counter may be removed but they must all be from the same pile. At the next turn, the player may elect to remove counters from another pile, or if it still exists, the same pile. The player picking up the last counter wins the game.

One of the most popular Nim games is the 3-4-5 game. It is played with 3 dimes, 4 nickels and 5 pennies. They are arranged in three horizontal rows as shown.

Dimes	000
Nickels	0000
Pennies	00000

An example of play might be:

Player A removes 2 dimes which leaves

Dimes   0
Nickels 0000
Pennies 00000

Player B removes 5 pennies which leaves

Dimes   0
Nickels 0000

Player A removes 3 nickels which leaves

Dimes   0
Nickels 0

Player B removes either the nickel or the dime but only one of them. The last coin was removed by Player A who is the winner. Did Player A win by chance or is there some method to his play? Of course, the answer is that Player A is a good gamester and knows that he can always win if he leaves an EVEN position, no matter what his opponent does, (providing he makes no errors). Remember the winning secret: *Always present your opponent with an EVEN position.*

To determine whether a position is EVEN or ODD, the numbers for each row are represented as binary numbers. The reader should refer to Table 11-1 if he is unfamiliar with the binary numbering system.

Table 11-1. Decimal/Binary Number Chart.

Decimal Number	Binary Number			
	2^3	2^2	2^1	2^0
0	0	0	0	0
1	0	0	0	1
2	0	0	1	0
3	0	0	1	1
4	0	1	0	0
5	0	1	0	1
6	0	1	1	0
7	0	1	1	1
8	1	0	0	0
9	1	0	0	1
10	1	0	1	0
11	1	0	1	1
12	1	1	0	0
13	1	1	0	1
14	1	1	1	0
15	1	1	1	1

The number of coins in each ROW is written in binary and each COLUMN is added. The position is EVEN if each column adds up to zero or an even number. Otherwise it is an ODD position. The player can always create an EVEN position from an ODD position, thus assuring a win. If the position is already EVEN when it is the player's turn to play, then he must make a *random* move and rely on his opponent to make a bad play.

To apply the above analysis to the 3-4-5 game mentioned earlier, we first assign binary values to the number of coins.

	COINS	NUMBER	BINARY VALUE
Dimes	000	3	011
Nickels	0000	4	100
Pennies	00000	5	101
			212

The first column adds up to 2, the second column is 1 and the third column is 2. The position is an ODD one. Player A creates an EVEN position by removing two dimes (Sum of the columns equals 202) thus assuring a WIN

	COINS	NUMBER	BINARY VALUE
Dimes	0	1	001
Nickels	0000	4	100
Pennies	00000	5	101
			202

The reader should explore other first moves (but in vain) in an attempt to find another move that would leave an EVEN position.

There is nothing mysterious about changing or converting decimal numbers to binary. As indicated in the Decimal/Binary Number Chart in Table 12-1, each binary position is a power of $2(2^0,2^1,2^2,2^3)$. A decimal number is simply a sum of the powers of $2(2^0 = 1, 2^1 = 2, 2^2 = 4, 2^3 = 8, 2^4 = 16$, etc.).

$$10 = 8+2 = 2^3+2^1 = 1010$$
$$6 = 4+2 = 2^2+2^1 = 0110$$

When a power of 2 is not represented in the conversion process, then a zero will appear in that position in the binary number. This can be noticed in the previous examples.

Consider putting 30 sticks in three piles, containing 9, 10, 11 sticks, as shown.

STICKS	NUMBER	BINARY VALUE
/////////	9	1001
//////////	10	1010
///////////	11	1011
		3022

Player A notes that the starting position is ODD, and by removing eight sticks from the last row, he can leave the position EVEN.

/////////	9	1001
//////////	10	1010
///	3	0011
		2022

Now assume that the opponent of Player A removes four sticks from the first row, thus leaving an ODD position.

/////	5	0101
//////////	10	1010
///	3	0011
		1122

Player A must choose his move so that the column totals leave an EVEN position. He accomplishes this by removing four sticks from row two.

/////	5	101
//////	6	110
///	3	011
		222

If Player B now removes three sticks from row 1, the new stick configuration is:

//	2	010
//////	6	110
///	3	011
		131

What must Player A do now to leave an EVEN position? Player A removes 5 sticks from row 2. Is this the only move that would leave an EVEN position? the result after this move:

//	2	010
/	1	001
///	3	011
		022

Player B is now getting tired of the game and removes 1 stick from row 1

/	1	001
/	1	001
///	3	011
		013

Player A now smiles, as he removes three sticks from the last row, knowing that he is guaranteed a win.

/	1	001
/	1	001
	0	000
		002

Player B mumbles something as he removes a stick from row 2. Player A removes the last stick from row 1 and of course wins the game.

Consider the example in Figure 11-1. If, at the beginning of a game of Nim, the coin configuration was as illustrated, would you rather play first or second? The second player could create an even position and be assured a win. Do you see why this is so?

COINS	NUMBER	BINARY VALUE
000	3	0011
0000	4	0100
000000000	9	1001
00000000000000	14	1110
		2222

Figure 11-1. The coins for a 3-4-9-14 game of nim.

If the sums of the binary columns indicated an ODD position, would you want to play first or second? Of course, the answer would be to play first if one wanted to insure victory.

11.2   The Four Magnetic Cores

Four magnetic cores are placed on a 36-celled square in such a manner that every one of the cores is in a straight line, with at least one of the other cores, yet no core is in line with another, as seen in Figure 11-2. The object of the game is to find the *maximum* number of ways the cores can be arranged in the squares.

Write a program to calculate the maximum arrangements of the magnetic cores. Do not count rotations or mirror reflections of a given arrangement. Did your program determine that 17 arrangements were possible?

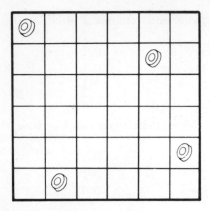

Figure 11-2. One arrangement of the four
magnetic core game.

11.3   Egg Game

Figure 11-3 illustrates a crate that can hold 36 eggs. It is divided into
36 compartments, 6 rows and 6 columns. How many eggs can be put
in the crate without having more than two eggs in any row, including
all the diagonal rows?

Two eggs are already placed in the crate at the start of the game.
Figure 11-3 illustrates the two positions of the eggs.

Write a program to play this game. Does the result computed by the
program compare with the arrangement shown in Figure 11-4?

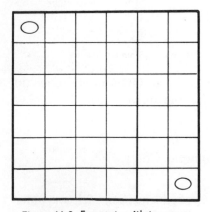

Figure 11-3. Egg crate with two eggs.

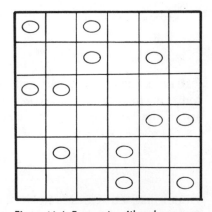

Figure 11-4. Egg crate with a dozen eggs.

11.4   The 50 Puzzle

The 6 by 6 array in Figure 11-5 contains 36 numbered squares. The
object of the puzzle is to connect any 3 consecutive squares either in a
vertical line, a horizontal line, or on diagonal lines, that add up to exactly
50. Typical connections that would be allowed are shown in Figure 11-6.

28	3	8	32	18	13
7	30	1	20	17	4
36	17	27	9	13	12
19	2	25	14	5	21
10	19	29	8	23	16
16	11	36	13	31	28

Figure 11-5. The 50 puzzle.

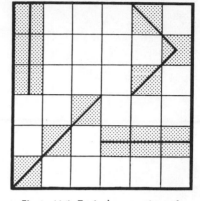

Figure 11-6. Typical connections of the 50 puzzle.

The flowchart in Figure 11-7 may be used to write a program to solve the 50 puzzle.

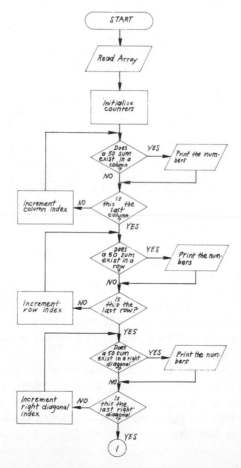

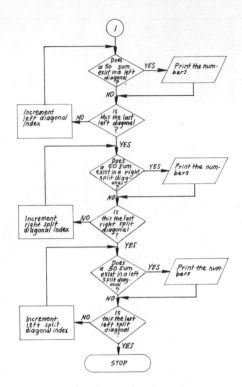

Figure 11-7. Flowchart of the 50 puzzle program.

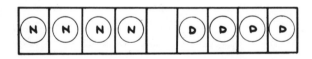

Figure 11-8. Board and starting position of the coins for the moving coin game.

11.5   Moving Coin Game

The *Moving Coin* game is played on the board shown in Figure 11-8. The board is divided into nine equal squares. At the start of the game, four nickles are placed on the four left-most squares and four dimes are placed on the right-most four squares. The center square is left vacant at the start of a game.

The object of the game is to move the coins to opposite ends of the board. Coins may move in two ways: they may jump over one of their own coins or one of the other coins and they may move one square at a time. However, the coins may not be moved backward. A solution to the game is as follows:

1.	M	D	L		9.	J	N	R		17.	J	N	R
2.	J	N	R		10.	M	N	R		18.	J	N	R
3.	M	N	R		11.	J	D	L		19.	M	D	L
4.	J	D	L		12.	J	D	L		20.	J	D	L
5.	J	D	L		13.	J	D	L		21.	J	D	L
6.	M	D	L		14.	J	D	L		22.	M	N	R
7.	J	N	R		15.	M	N	R		23.	J	N	R
8.	J	N	R		16.	J	N	R		24.	M	D	L

The reader is advised to try the game prior to writing a program to do so.

11.6 Pentomino Game

A *pentomino* is a plane figure formed by five contiguous equal squares. There are 12 possible ways to arrange five squares in this manner; therefore, there are 12 different pentominoes. The pentomino game is played by arranging the 12 pentominoes, shown in Figure 11-9, into a 6 by 10 squared rectangular box. Although there are over 2000 ways of accomplishing this, it is not as easy as it sounds. The reader is advised to make a set of pentominoes out of cardboard and attempt it prior to writing a program.

Figure 11-10 illustrates several arrangements of the 12 pentominoes.

11.7 One Pile Pickup

This game is a slight variation of Nim and uses only one pile of articles. Each player is required to pick up at least one but less than k articles. The *loser* of the game is the person picking up the last article.

Consider the following game, where two players start with 9 checkers. Each player must not pick up more than two checkers at any one time; therefore, $k = 3$.

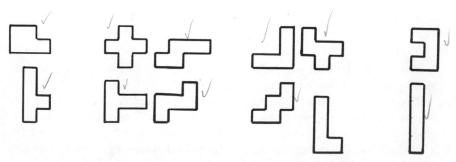

Figure 11-9. The 12 pentominoes.

Figure 11-10. Eight pentomino arrangements of size 6 by 10.

Starting Pile of Checkers	000000000
Player 1 takes 2 checkers leaving 7	0000000
Player 2 takes 1 checker leaving 6	000000
Player 1 takes 1 checker leaving 5	00000
Player 2 takes 1 checker leaving 4	0000
Player 1 takes 1 checker leaving 3	000
Player 2 takes 1 checker leaving 2	00
Player 1 wins the game by taking	0
1 checker and leaving 1 checker	
for Player 2 to take.	

In general, the first player able to leave $nk + 1$ checkers can win by leaving $n(k-1) + 1$ checkers at the next play.

Player 1 in the illustrated game used this strategy when he took two checkers on the first move ($nk+1 = 2 \cdot 3 + 1 = 7$ checkers), and again when he left five checkers after his second move ($n(k-1)+1 = 2(3-1)+1 = 5$), three checkers after his third move ($1(3-1)+1 = 3$ checkers), and 1 checker after his last move ($0(3-1)+1 = 1$ checker).

11.8   Marked Squares Game

This game is played on a board with 9 rows and 9 columns. Players alternate in marking a square at a time, each using his own distinguishing mark. Whenever a player marks the last square in a horizontal, vertical or diagonal line, he is credited with all the squares in this line. Each square is scored as one point. The player with the higher final score wins the game.

11.9   Sliding Box Puzzle

Martin Gardner in the March 1965 issue of *Scientfic American* introduced the Sliding Box Puzzle of Figure 11-11. The object of the puzzle was to change the pattern on the left to the pattern on the right. The number squares move freely within the box and up until 1965 the best solution on record required 36 moves. This method employed the following moves:

125431237612376123754812365765847856.

Computers were used in the solution to the puzzle and they found 634 methods that required less than 36 moves.

10 methods requiring 30 moves
112 methods requiring 32 moves
512 methods requiring 34 moves

The 10 minimum-move methods that were computed by a computer took about 2½ minutes. These methods are listed below:

34785217437486386521478652147 8
125874312587431631526528741256

34785217852178564385643642145 8
145875365341653412874128741256

34521543547821478638621475865 8
143142587316312587125465487456

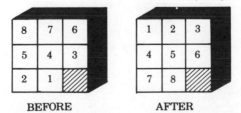

BEFORE   AFTER

Figure 11-11. The sliding box puzzle.

34521576435768217684356842 1456
34587513465132871324653 2487456

125874852831825743163125741258
14785247863865247 1861741521478

Write a program that will solve this puzzle.

PART THREE

Game Playing Topics

CHAPTER 12

Game Playing With BASIC

12.1   Introduction to BASIC

EVER SINCE THE FIRST COMPUTER was invented, computer manufacturers, programmers, and computer users have been trying to find ways to simplify programming. Their aim has been to develop programming languages that are easy to understand, easy to learn, easy to use, and that are applicable to a large number of computers.

FORTRAN, an acronym derived from FORmula TRANslation, is a scientifically oriented programming language first developed in 1957. It has a great deal of machine independence and is probably the most widely used compiler language in the world, partly because FORTRAN compilers have been developed for most computers now in use. The game programs in Part I of this book were all written in FORTRAN.

Another language called BASIC   Beginner's All-purpose Symbolic Instruction Code) is also very easy to learn and is an ideal language for programming game problems. This interactive language, first developed in the late 1960's, has subsequently been implemented on many computer systems, including minicomputer systems, and is used as a language by many commercial time-sharing service companies.

The BASIC programs in this chapter were executed on a General Electric time-sharing system, simply because the author had access to G. E. computer time-sharing services and not because of any special preference. The reader may also be interested in reading another book by the author entitled *Game Playing with BASIC*, Hayden Book Company, 1976.

12.2   Introduction to Time-Sharing

TIME-SHARING in its present context is a reasonably new computer technique and has been of practical use for only a few years. Time-sharing is essentially a technique to make use of large computers economically by sharing a computer system among several or many users simultaneously.

In a typical time-sharing computer system, several users may communicate with a remotely located Central Computer by means of individual input/output stations. An input/output (I/O) station usually consists of a teletypewriter (such as a Model 33 or Model 35 Teletype unit), a visual display unit, or a typewriter. These input/output stations are used to prepare programs, to input programs to the Central Computer and to accept, print or display computed results from the Central Computer.

Communications between the input/output stations and the remotely located Central Computer is via common telephone lines. Figure 12-1 illustrates a typical time-sharing configuration with a Central Computer System located in a centralized area and input/output stations located in several remote areas. A user in an input/output station initiates actions which connect him to the computer, and from then on the user can communicate directly with the computer system.

The number of input/output stations that can be connected to the computer is a function of a specific system; however, many systems have provisions for around 40 stations and a few larger systems will allow 200 or 300 users to work simultaneously with a Central Computer System.

To understand time-sharing and how this type of system may be used for game playing, let us first look at a specific system and an associated programming language. The language and system described is the General Electric time-sharing system and the BASIC programming language. This system is representative of many time-sharing systems in that the

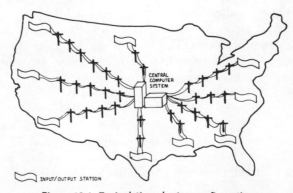

INPUT/OUTPUT STATION

Figure 12-1. Typical time-sharing configuration.

time-sharing hardware is present and BASIC is a language that is similar to many other time-sharing conversional languages. The equipment used in the GE system consists of a computer system, a data communications processor and a disc storage unit. The input/output stations are Model 33 Teletype units. The Model 33 Teletype unit has, in addition to a keyboard and roll paper output mechanism, a paper tape reader and punch. The Teletype unit may be used off-line from the computer system to prepare a punched paper tape of program instructions and data. The Teletype unit is used to gain access to the system. Briefly, what takes place is this. The user identifies himself by typing the word HELLO on the keyboard. This initiates a short series of questions and answers which identifies the user and the problem. Specifically, the user supplies a user number and the name of the programming language which he will use. He specifies whether the problem about to be named is NEW or OLD, and he types the problem name. If the problem is OLD, the system retrieves it from the disc storage area, and he may modify or use it in any manner he chooses. If the problem is NEW, the user composes, edits and tests his program right at the Model 33 teletype keyboard. When the program is ready to be executed, the user types the word, RUN. The computer system immediately performs its calculations and responds by typing its reply on the user's teletype unit.

Using a Basic Program. A program that simulates the operation of a slot machine is used to show the reader how a user communicates with a program in a time-sharing system. A sample run of the slot machine BASIC program appears in Figure 12-2.

The previous *conversation* between an operator of a teletypewriter terminal and a remotely located time-sharing computer system illustrates the ease of using this type of system. In this example, all underlined entries are keyed in by the operator at the terminal. All other figures come from the time-sharing computer.

1. After the operator gains control of the computer, the computer causes USER NUMBER to be typed. The operator responds by typing his identification number.
2. The computer types SYSTEM to ask what programming language the operator wishes to use. The operator replies by typing BASIC.
3. The computer causes NEW or OLD to be typed. NEW refers to placing a new computer program in memory and OLD refers to a program that has been previously prepared and is currently stored on the computer's disc storage unit. The operator types OLD.
4. The computer causes OLD PROBLEM NAME to be printed. The operator types the name of the slot machine program, SLOT.
5. The computer types out WAIT, and after a few seconds, it types out READY.

```
USER NUMBER--1863W
SYSTEM--BASIC
NEW OR OLD--OLD
OLD PROBLEM NAME--SLOT
WAIT.

READY

RUN

SLOT  12:08  CH  TUE  12/06/67

THIS PROGRAM SIMULATES THE OPERATION OF A SLOT MACHINE. YOU
SUPPLY A NUMBER BETWEEN 1 AND 500 AND THE AMOUNT OF MONEY
YOU WANT TO START WITH. THE MACHINE WILL ACCEPT ANY NUMBER OF
SILVER DOLLARS ON EACH PLAY. GOOD LUCK.

WHAT NUMBER BETWEEN 1 AND 500 DO YOU CHOOSE?  37

HOW MUCH MONEY DO YOU WANT TO START WITH?  2

CHERRY    CHERRY    PLUM

A WINNER. YOU NOW HAVE 6 DOLLARS.

TYPE 222 IF YOU WANT TO TRY AGAIN?  222

JACKPOT JACKPOT JACKPOT

A JACKPOT WINNER. YOU NOW HAVE 600 DOLLARS.

TYPE 222 IF YOU WANT TO TRY AGAIN?  222

CHERRY    ORANGE    BELL

A LOSER. YOU HAVE LOST ALL YOUR MONEY. BETTER LUCK NEXT TIME.

TIME    7 SEC.

BYE

***OFF AT 12:15  CH  TUE  12/06/67
```

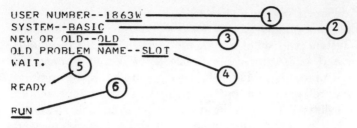
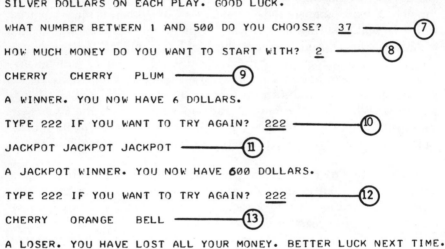

Figure 12-2. Sample run of slot machine program.

6. The operator types RUN which informs the computer that he wants to run the program SLOT.

7. Computer executes the program SLOT which causes a message and a question to be typed. The operator is asked to type a number between 1 and 500. The computer uses this number to start generating random numbers which are used to determine the three slot machine values. The operator answers this question by typing 37.

8. The computer causes another question to be typed. This question

refers to the amount of money the operator wants to bet. The operator types 2 which represents 2 silver dollars.

9. The computer types the slot machine payout line of CHERRY, CHERRY, PLUM and a message indicating that the operator won six dollars.

10. The computer types a message that allows the operator to stop or continue playing. If the operator types 222, the game will continue. If the operator types any other number, the computer will stop playing the game and print the message "YOUR GAME IS OVER." In this example, the operator desires to continue and he types 222.

11. The computer types another slot machine payoff line. This payoff was a jackpot and the operator now has won 600 dollars.

12. The operator types 222 to indicate to the computer that he wants to play again.

13. The computer types the payoff line CHERRY, ORANGE, BELL and a message indicating that this is a losing combination of symbols.

14. The computer types the amount of computer time that was used while playing the slot machine program.

15. The operator types BYE (he could also type GOODBYE) to indicate that he wishes to disconnect the teletype unit from the computer system.

16. The computer types a sign-off message which indicates the time and date that program execution terminated.

A time-sharing computer system is a convenient way of solving many scientific, business and game playing problems. This conversational mode of communicating with the computer is an ideal way to conduct computerized gaming.

12.3 Game Programs in the GE Library

Several game programs are stored in the GE time-sharing library. They are used primarily as demonstration programs and are available for play by any user of the system. The programs are written in BASIC and ALGOL programming languages and include programs to play *Blackjack, Tic-Tac-Toe, Battle of Numbers, Slot Machine* and *Craps*, as well as many other games.

Sample runs of four of these programs are as follows.

Tic-Tac-Toe. This program will play Tic-Tac-Toe with an operator. It keeps track of the operator's moves and prints the board configuration every time the computer makes its move. An operator informs the computer of his move by indicating the Row and Column that the desired position lies in. The following diagram illustrates this procedure.

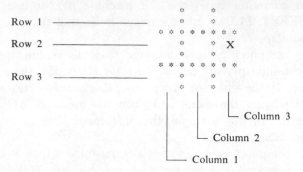

For example, the position marked by an X is in Row 2 and Column 3. Input to the Tic-Tac-Toe program would be represented as 2,3.

The human operator always starts the game. In the examples of Figure 12-3, the operator's moves are indicated by *YOU* and the computer's moves are marked by *235*. The operator wins the first game and the computer wins the second game.

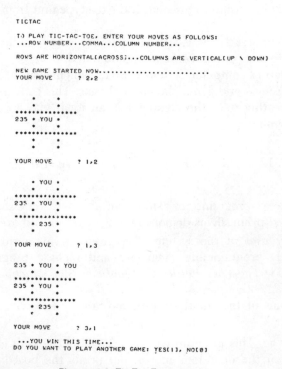

Figure 12-3. Tic-Tac-Toe printouts.

```
NEW GAME STARTED NOW.........................
YOUR MOVE      ? 3,1


     .    .
     .    .
.................
   . 235 .
.................
YOU .    .
     .    .

YOUR MOVE      ? 2,1

235 .    .
.................
YOU . 235 .
     .    .
.................
YOU .    .
     .    .

YOUR MOVE      ? 1,2

 ... THE 235 WINS THIS TIME...

235 . YOU .
     .    .
.................
YOU . 235 .
     .    .
.................
YOU .    . 235
     .    .

DO YOU WANT TO PLAY ANOTHER GAME: YES[1], NO[0]          ? 0
ITS BEEN FUN, CALL ME AGAIN SOMETIME
```

Figure 12-3. Tic-Tac-Toe printouts (cont'd).

```
THIS PROGRAM PLAYS THE GAME OF 'CRAPS' WITH YOU....


WHAT NUMBER BETWEEN 1 AND 711 IS LUCKY FOR YOU TODAY? 34
SPLENDID.....YOU ARE GIVEN  70    DOLLARS TO PLAY WITH.


YOU ROLL FIRST....


HOW MUCH DO YOU BET? 20

YOU ROLL   5     AND   3    SO YOUR POINT IS 8
YOU ROLL   2     AND   4    ...ROLL AGAIN.
YOU ROLL   3     AND   2    ...ROLL AGAIN.
YOU ROLL   2     AND   1    ...ROLL AGAIN.
YOU ROLL   3     AND   1    ...ROLL AGAIN.
YOU ROLL   1     AND   3    ...ROLL AGAIN.
YOU ROLL   3     AND   6    ...ROLL AGAIN.
YOU ROLL   2     AND   4    ...ROLL AGAIN.
YOU ROLL   5     AND   3    AND MAKE YOUR POINT

YOU NOW HAVE  90    DOLLARS
HOW MUCH DO YOU BET? 90

YOU ROLL   6     AND   4    SO YOUR POINT IS 10
YOU ROLL   5     AND   4    ...ROLL AGAIN.
YOU ROLL   3     AND   1    ...ROLL AGAIN.
YOU ROLL   6     AND   5    ...ROLL AGAIN.
YOU ROLL   1     AND   4    ...ROLL AGAIN.
YOU ROLL   3     AND   4    AND LOSE...


YOU HAVE RUN OUT OF MONEY....SORRY ABOUT THAT.
```

Figure 12-4. Printout for craps.

Craps. This program will play a simple game of *Craps* with an operator. The program asks the operator to supply a number between 1 and
711. This number is used to select an amount of money that the operator is to play with. The printout of Figure 12-4 illustrates two plays of this
program. In the first example the operator bets $20.00 and makes his
point of 8. In the second example the operator bets and loses $90.00
as he did not make his point of 10 before he rolled a 7.

Blackjack. Blackjack is the most popular card game found in casinos
throughout the world. It is also a popular demonstration game at computer conferences and other gatherings where computers are being
demonstrated to the public. The blackjack printouts in Figure 12-5 were
generated by a teletype unit located in Daytona Beach, Florida and a
GE-265 computer located in New York City. The blackjack game is
played by 1) gaining control of the computer system in New York;
2) informing the system that you want the blackjack rules to be typed
(type 21 INFO to accomplish this); 3) informing the system that you
want to play with the blackjack program (type BLKJAK to accomplish
this); 4) informing the system of the time of day [(in this example the
time was 1730 (5:30)]. This time value is used by the program to generate a random point to start card values; and 5) informing the system
of your wager ($50.00 was bet on the first hand, $30.00 on the second,
$30.00 on the third and $2.00 on the last). The blackjack program then
prints the dealer's *up card* and the *player's two cards*. In the first example,
the player bets $50.00 and has a standoff since both the player and com-
puter had a total card count of 20. In the second example, the player
bets $30.00 and wins; in the third example, the player bets $30.00 and
again has a standoff; and in the last example the player bets and loses
$2.00 since the computer's card total of 19 is closer to 21 than the player's
total of 13.

Battle of Numbers. This game is a special case of the game of *Nim*.
The player must determine the number of objects in a pile, whether
taking the last object results in a win or loss, who plays first, and the
minimum and maximum number of objects that the computer or player
can remove from the pile at any one time. In the game of Figure 12-6 the
player chooses 76 objects, that the one who is forced to take the last
object loses the game, that the computer plays first and that only 2, 3,
4, 5 or 6 coins can be removed at any one time. The computer wins the
game by forcing the player to remove the last object from the pile. If
the reader is interested in how the computer determines a move, he
should study the game of Nim as discussed in Chapter 11.

21INFO

PLAY THE GAME OF BLACKJACK USING THE EXACT RULES AS FOUND
IN MOST CASINOS IN LAS VEGAS.
                    HOW TO PLAY THE GAME

     WHEN I SAY ''THIS DEALER GETS A BREAK AT 1945, WHAT TIME IS IT
NOW?'', TELL ME, AND THEN DEPRESS THE RETURN KEY. NEXT I WILL
ASK THE AMOUNT OF YOUR WAGER. TYPE A NUMBER FROM 0 TO 500, AND THEN
DEPRESS THE RETURN KEY. WHEN YOU WISH TO TERMINATE THE GAME MAKE A
WAGER OF 0 OR LESS.
     I WILL DEAL MYSELF 2 CARDS, SHOW YOU ONE, THEN DEAL YOU 2 CARDS,
AND ASK IF YOU WANT A ''HIT''. AT THIS POINT YOU HAVE SEVERAL OPTIONS
DEPENDING ON THE HOLE CARDS YOU HOLD AND MY UP CARD. YOU MAY STICK
BY TYPING A 0, TAKE A HIT BY TYPING A 1, GO DOWN FOR DOUBLES BY
TYPING A 2, OR YOU MAY SPLIT A PAIR BY TYPING A 3.
     IF MY UP CARD IS AN ACE I WILL ASK IF YOU WANT ''INSURANCE''
IF YOU TAKE INSURANCE [BY TYPING A 1]THIS MEANS THAT YOU ARE BETTING
1/2 OF YOUR WAGER THAT I DO HAVE BLACKJACK. IF I DO, I PAY YOU
2 TO 1 ON YOUR INSURANCE BET AND YOU LOSE YOUR ORIGINAL BET SINCE
I DO HAVE BLACKJACK, WITH THE END RESULT BEING THAT YOU ARE
EVEN FOR THE HAND. IF I DO NOT HAVE BLACKJACK YOU LOSE YOUR
INSURANCE BET AND THE GAME CONTINUES FROM THERE. IF YOU REFUSE
INSURANCE BY TYPING A 0 THE GAME CONTINUES AS NORMAL.
     WHEN YOU FINALLY STICK, BY TYPING A 0, I WILL DRAW CARDS UNTIL
I HAVE AT LEAST A HARD 17 [HARD MEANING THE TOTAL DOES NOT INCLUDE
AN ACE BEING COUNTED AS 11] OR A SOFT 18 [TOTAL INCLUDES AN ACE
COUNTED AS 11].
     I PAY 1.5 TO 1 ON BLACKJACK, IT IS LEGAL TO DOUBLE DOWN
ON A SPLIT HAND, YOU DO NOT LOSE ON A TIE HAND, AND I DO NOT
RECOGNIZE 5 CARDS AND UNDER 21.
     ****GOOD LUCK****
FOR A DEFINITION OF VARIABLES USED IN PLA*21, LIST THIS
PROGRAM FROM LINE NUMBER 100 ON.

BLKJAK

RUN PROGRAM NAMED [21INFO***] FOR INFORMATION ON PLAYING THIS GAME

THIS DEALER GETS A BREAK AT 1945, WHAT TIME IS IT NOW? 1730

WAGER? 50

I SHOW   10   OF CLUBS
FIRST CARD IS   QUEEN OF HEARTS
NEXT   CARD IS   JACK OF SPADES
HIT? 0
YOUR TOTAL IS   20
MY HOLE CARD IS   JACK OF HEARTS
MY TOTAL IS 20
YOU'RE EVEN

WAGER? 30

I SHOW   KING OF DIAMONDS
FIRST CARD IS   3   OF HEARTS
NEXT   CARD IS   KING OF CLUBS
HIT? 1
NEXT   CARD IS   ACE OF DIAMONDS
HIT? 1
NEXT   CARD IS   6   OF DIAMONDS
HIT? 0
YOUR TOTAL IS   20
MY HOLE CARD IS   9   OF DIAMONDS
MY TOTAL IS 19
YOU'RE AHEAD $ 30

Figure 12-5. Blackjack printouts.

WAGER? 30

```
I SHOW                           8      OF SPADES
FIRST CARD IS    QUEEN OF DIAMONDS
NEXT  CARD IS    4      OF CLUBS
HIT? 0
YOUR TOTAL IS    14
MY HOLE CARD IS                  7      OF HEARTS
I DRAW                           4      OF HEARTS
MY TOTAL IS 19
YOU'RE EVEN
```

WAGER? 2

```
I SHOW                           9      OF HEARTS
FIRST CARD IS    4      OF DIAMONDS
NEXT  CARD IS    9      OF SPADES
HIT? 0
YOUR TOTAL IS    13
MY HOLE CARD IS                  JACK OF DIAMONDS
MY TOTAL IS 19
YOU'RE BEHIND $ 2
```

Figure 12-5. Blackjack printouts (cont'd).

BATNUM

```
THIS PROGRAM PLAYS 'THE BATTLE OF NUMBERS.'
THE GAME IS PLAYED WITH A PILE OF OBJECTS, SOME OF
WHICH ARE REMOVED ALTERNATELY BY YOU AND THE MACHINE.
YOU MUST SPECIFY WHETHER WINNING IS DEFINED AS TAKING
OR NOT TAKING THE LAST OBJECT, THE ORIGINAL NUMBER OF
OBJECTS IN THE PILE, WHO GOES FIRST, AND THE MINIMUM
AND MAXIMUM NUMBER OF OBJECTS WHICH CAN BE REMOVED AT
ONE TIME.  TYPING '0' FOR YOUR MOVE WILL CAUSE A
FORFEIT, AND TYPING '0' FOR THE PILE SIZE WILL CAUSE
THE TERMINATION OF THE GAME.
ENTER PILE SIZE: ? 76
ENTER WIN OPTION - 1 TO TAKE LAST, 2 TO AVOID LAST: ? 2
ENTER MIN AND MAX: ? 2,6
ENTER START OPTION - 1 MACHINE FIRST, 2 YOU FIRST: ? 1
MACHINE TAKES 3      AND LEAVES 73
YOUR MOVE: ? 5
MACHINE TAKES 3      AND LEAVES 65
YOUR MOVE: ? 6
MACHINE TAKES 2      AND LEAVES 57
YOUR MOVE: ? 6
MACHINE TAKES 2      AND LEAVES 49
YOUR MOVE: ? 6
MACHINE TAKES 2      AND LEAVES 41
YOUR MOVE: ? 4
MACHINE TAKES 4      AND LEAVES 33
YOUR MOVE: ? 6
MACHINE TAKES 2      AND LEAVES 25
YOUR MOVE: ? 5
MACHINE TAKES 3      AND LEAVES 17
YOUR MOVE: ? 4
MACHINE TAKES 4      AND LEAVES 9
YOUR MOVE: ? 2
MACHINE TAKES 6      AND LEAVES 1
YOUR MOVE: ? 1
YOU LOSE
ENTER PILE SIZE: ? 0
```

Figure 12-6. Battle of numbers printout.

12.4   A BASIC Sieve of Eratosthenes Program

The BASIC program of Figure 12-7 uses the *Sieve of Eratosthenes* to generate the first 79 prime numbers. A flowchart of this program is shown in Figure 12-8.

```
10   REM SIEVE OF ERATOSTHENES PROGRAM
20   DIM P(400)
30   REM FILL ARRAY P WITH THE NUMBERS 1 THRU 400
40   FOR X = 1 TO 400
50   LET P(X) = X
60   NEXT X
70   REM CLEAR EVEN NUMBERS STARTING WITH 4
80   FOR Y = 4 TO 400 STEP 2
90   LET P(Y) = 0
100  NEXT Y
110  FOR Z = 3 TO SQR(400) STEP 2
120  REM SKIP IF P(Z) CELL CONTAINS A ZERO
130  IF P(Z) = 0 THEN 180
140  REM CLEAR ALL NUMBERS THAT ARE NOT PRIME
150  FOR D = 2 * Z TO 400 STEP Z
160  LET P(D) = 0
170  NEXT D
180  NEXT Z
190  PRINT " THE FIRST 79 PRIME NUMBERS "
200  PRINT " CALCULATED BY THE SIEVE OF ERATOSTHENES "
210  PRINT
220  PRINT
230  REM PRINT ARRAY P CONTAINING THE PRIME NUMBERS
240  FOR J = 1 TO 400
250  PRINT P(J);
260  NEXT J
270  END
```

Figure 12-7. Program for the Sieve of Eratosthenes.

Output of this program is shown in Figure 12-9. The prime numbers less than 400 are 1, 2, 3, 5, 7, 11, 13, 17, . . . , 383, 389 and 397.

12.5   A BASIC Geometric Magic Square Generating Program

A *Geometric Magic Square* is an array of numbers where the *product* of the numbers in every column, row and main diagonal is the same. Each number of the square is represented by a base value and an exponent. The base value remains the same in all the positions of the square and the exponent values are the numbers in an ordinary odd order magic square. For example, an Order 3 Geometric Magic Square with a base of 2 would appear as shown in Figure 12-10.

A flowchart of a Geometric Magic Square Generating Program is shown in Figure 12-11. The BASIC program is shown in Figure 12-12.

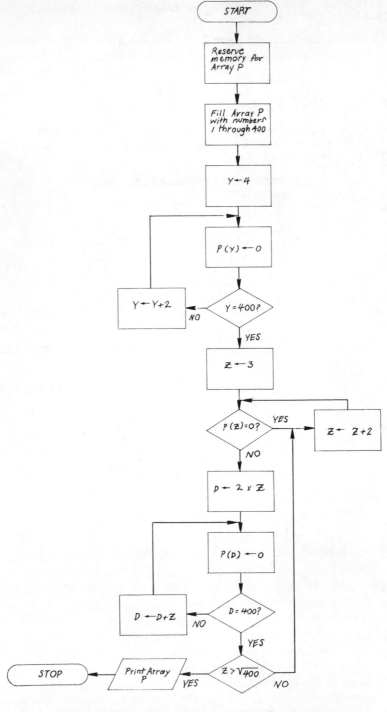

Figure 12-8. Flowchart of the Sieve of Eratosthenes program.

THE FIRST 79 PRIME NUMBERS
CALCULATED BY THE SIEVE OF ERATOSTHENES

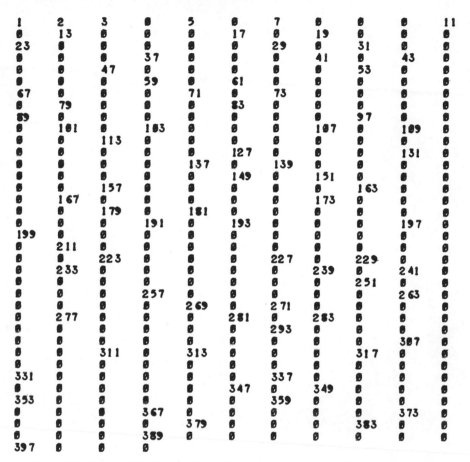

Figure 12-9. Output of the Sieve of Eratosthenes program (prime numbers less than 400).

2^8	2^1	2^6		256	2	64
2^3	2^5	2^7	$=$	8	32	128
2^4	2^9	2^2		16	512	4

Geometric Magic
Square using base
and exponent values.

Geometric Magic
Square using
integer values.

Figure 12-10. An order 3 geometric magic square.

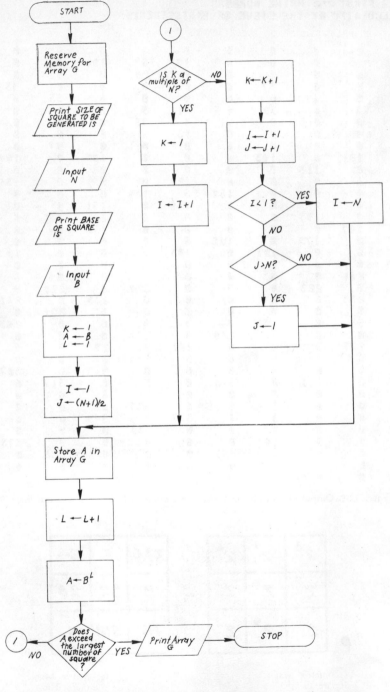

Figure 12-11. Flowchart of the geometric magic square generating program.

```
10   REM GEOMETRIC MAGIC SQUARE GENERATING PROGRAM
20   REM GENERATES AN ODD GEOMETRIC MAGIC SQUARE OF SIZE N BY N
30   DIM G(21,21)
40  PRINT " SIZE OF SQUARE TO BE GENERATED IS ";
50   INPUT N
60  PRINT " BASE OF SQUARE IS ";
70   INPUT B
80   LET K = 1
90   LET A = B
100  LET L = 1
110  LET I = 1
120  LET J = ( N + 1 ) / 2
130  LET G(I,J) = A
140  LET L = L + 1
150  LET A = B↑L
160  IF A > (B↑(N↑2)) THEN 295
170  IF K < N THEN 210
180  LET K = 1
190  LET I = I + 1
200  GO TO 130
210  LET K = K + 1
220  LET I = I - 1
230  LET J = J + 1
240  IF I <> 0  THEN 270
250  LET I = N
260  GO TO 130
270  IF J <= N THEN 130
280  LET J = 1
290  GO TO 130
295  PRINT
296  PRINT
300  PRINT N " BY " N " GEOMETRIC MAGIC SQUARE"
310  PRINT
320  FOR I = 1 TO N
330  FOR J = 1 TO N
340  PRINT G(I,J);
350  NEXT J
360  PRINT
370  PRINT
380  PRINT
390  NEXT I
400  END
```

Figure 12-12. Geometric magic square generating program.

This program types the messages

SIZE OF SQUARE TO BE GENERATED IS?

and

BASE OF SQUARE IS?

When the operator answers these two questions, the program will generate the requested magic square. For example, assume that you wanted the program to generate an Order 3 square using a base of 3. You would answer the first question by typing 3 and the second question by typing another 3. The computer would cause the top square of Figure 12-13 to be typed.

Figure 12-13 also illustrates two other Order 3 Geometric Magic Squares. One of the squares uses a base of 4 while the other one has a base of 5.

6561	3	729
27	243	2187
81	19683	9

65536	4	4096
64	1024	16384
256	262144	16

390625	5	15625
125	3125	78125
625	1953125	25

Figure 12-13. Three 3 by 3 geometric magic squares.

12.6   A BASIC 15 Puzzle Program

The BASIC program of Figure 12-14 will solve the 15 Puzzle that was described in Chapter 5.

```
5     REM 15 PUZZLE PROGRAM
10    REM THIS PROGRAM WILL DETERMINE IF A GIVEN NUMBER
20    REM ARRANGEMENT IS POSSIBLE OR IMPOSSIBLE
30    REM PRINT GAME DESCRIPTION
40    PRINT "                    15 PUZZLE "
50    PRINT
60    PRINT "   THE 15 PUZZLE WAS INVENTED BY SAM LOYD IN 1878. IT "
70    PRINT "   HAS BEEN AN EXTREMELY POPULAR PUZZLE IN EUROPE AND "
80    PRINT " AMERICA. THE 15 PUZZLE CONSISTS OF A SQUARE NUMBER "
90    PRINT " ARRANGEMENT WITH THE NUMBERS 1 TO 15 AND A BLANK. THE "
100   PRINT " BLANK IS REPRESENTED BY THE NUMBER 16. ANY ONE OF THE "
110   PRINT " NUMBERS TO THE IMMEDIATE RIGHT, LEFT, TOP OR BOTTOM OF "
120   PRINT " THE BLANK SQUARE CAN BE MOVED INTO THE BLANK POSITION. "
130   PRINT " THE OBJECT OF THE PUZZLE IS TO START WITH THE NUMBER "
140   PRINT " CONFIGURATION "
150   PRINT "            1     2     3     4 "
160   PRINT
170   PRINT "            5     6     7     8 "
180   PRINT
190   PRINT "            9    10    11    12 "
200   PRINT
210   PRINT "           13    14    15    16 "
220   PRINT
230   PRINT " AND FINISH WITH A DIFFERENT NUMBER ARRANGEMENT SAY "
240   PRINT " THE ONE SHOWN BELOW "
250   PRINT
260   PRINT "            4     8    12    14 "
270   PRINT
280   PRINT "            3     7    11    16 "
290   PRINT
300   PRINT "            2     6    10    15 "
310   PRINT
320   PRINT "            1     5     9    13 "
330   PRINT
340   PRINT " THERE IS ONE SLIGHT CATCH TO THE PUZZLE --- THERE "
350   PRINT " ARE 10,461,394,944,000 NUMBER ARRANGEMENTS THAT "
360   PRINT " ARE IMPOSSIBLE TO OBTAIN. THERE ARE ALSO  THE SAME "
370   PRINT " NUMBER OF POSSIBLE ARRANGEMENTS. "
```

```
380 PRINT
390 PRINT " -------------------------------------------------------"
400 PRINT " --- RULES FOR USING THE 15 PUZZLE PROGRAM --- "
410 PRINT
420 PRINT " AFTER THE PROGRAM TYPES A DESCRIPTION AND RULES OF THE "
430 PRINT " 15 PUZZLE THE FOLLOWING MESSAGE WILL BE PRINTED "
440 PRINT
450 PRINT "    TYPE THE NUMBER ARRANGEMENT TO BE ACHIEVED "
460 PRINT "    IN ROW ORDER ( FIRST ROW FIRST, SECOND ROW "
470 PRINT "    NEXT, ETC.). SEPARATE EACH NUMBER BY A "
480 PRINT "    COMMA. THE BLANK SQUARE IS REPRESENTED BY 16."
490 PRINT
500 PRINT " AFTER YOU TYPE THE NUMBER ARRANGEMENT TO BE "
510 PRINT " ACHIEVED AND PRESS RETURN THE PROGRAM WILL "
520 PRINT " DETERMINE IF THE ARRANGEMENT CAN OR CANNOT BE "
530 PRINT " ACHIEVED AND WILL PRINT THE NUMBER ARRANGEMENT "
540 PRINT " AND THE APPROPRIATE OF THE FOLLOWING TWO MESSAGES "
550 PRINT
560 PRINT "    THE FOLLOWING NUMBER ARRANGEMENT "
570 PRINT "    IS IMPOSSIBLE TO ACHIEVE "
580 PRINT
590 PRINT "    THE FOLLOWING NUMBER ARRANGEMENT "
600 PRINT "    IS POSSIBLE TO ACHIEVE"
610 PRINT
620 PRINT " YOU MAY INPUT ANOTHER NUMBER ARRANGEMENT BY"
630 PRINT " TYPING 666 WHEN REQUESTED TO DO SO . YOU MAY "
640 PRINT " STOP BY TYPING 777 WHEN REQUESTED TO DO SO. "
642 DIM S(4,4)
650 PRINT
660 PRINT
670 PRINT " TYPE THE NUMBER ARRANGEMENT TO BE ACHIEVED"
680 PRINT " IN ROW ORDER ( FIRST ROW FIRST, SECOND ROW"
690 PRINT " NEXT, ETC.). SEPARATE EACH NUMBER BY A "
700 PRINT " COMMA. THE BLANK SQUARE IS REPRESENTED BY 16."
705 PRINT
710 INPUT A,B,C,D,E,F,G,H,I,J,K,L,M,N,O,P
720 LET S(1,1) = A
730 LET S(1,2) = B
740 LET S(1,3) = C
750 LET S(1,4) = D
760 LET S(2,1) = E
770 LET S(2,2) = F
780 LET S(2,3) = G
790 LET S(2,4) = H
800 LET S(3,1) = I
810 LET S(3,2) = J
820 LET S(3,3) = K
830 LET S(3,4) = L
840 LET S(4,1) = M
850 LET S(4,2) = N
860 LET S(4,3) = O
870 LET S(4,4) = P
871 PRINT
872 PRINT A;B;C;D
873 PRINT
874 PRINT E;F;G;H
875 PRINT
876 PRINT I;J;K;L
877 PRINT
878 PRINT M;N;O;P
880 LET T = 0
890 IF S(1,2) = 16 THEN 980
900 IF S(1,4) = 16 THEN 980
910 IF S(2,1) = 16 THEN 980
920 IF S(2,3) = 16 THEN 980
930 IF S(3,2) = 16 THEN 980
940 IF S(3,4) = 16 THEN 980
950 IF S(4,1) = 16 THEN 980
960 IF S(4,3) = 16 THEN 980
```

```
970 GO TO 990
980 LET T = T+1
990 FOR I = 1 TO 4
1000FOR J = 1 TO 4
1010FOR K = 1 TO 4
1020FOR L = 1 TO 4
1030IF S(I,J) > S(K,L) THEN 1070
1040NEXT L
1050NEXT K
1060GO TO 1080
1070LET T = T + 1
1080LET S(I,J) = 16
1082NEXT J
1084NEXT I
1090IF T - 2 * INT(T / 2) = 0 THEN 1140
1100PRINT
1110PRINT
1120PRINT" THE ABOVE 15 PUZZLE NUMBER ARRANGEMENT CANNOT BE ACHIEVED"
1130GO TO 1170
1140PRINT
1150PRINT
1160PRINT" THE ABOVE 15 PUZZLE NUMBER ARRANGEMENT CAN BE ACHIEVED"
1170PRINT
1180PRINT
1190PRINT " TYPE 777 IF YOU WANT TO STOP."
1200PRINT " TYPE 666 IF YOU WANT TO CONTINUE."
1210INPUT X
1220IF X = 777 THEN 1270
1230IF X = 666 THEN 650
1240PRINT
1250PRINT " CANT YOU READ ---- TYPE EITHER 777 OR 666."
1260GO TO 1210
1270END
```

Figure 12-14. BASIC 15 puzzle program.

This program causes the teletype unit to type a short description of the 15 Puzzle and rules for using the program. After the playing rules are typed, the program causes a message to be typed that directs the operator to type the number arrangement that the program will work with. After the operator types the 16 required entries, the program prints the array of numbers and a statement indicating that the chosen number arrangement can or cannot be achieved. The operator can continue playing this program by typing 666 when requested to do so or can stop program execution by typing 777. Any other number will cause the message "CANT YOU READ—TYPE EITHER 777 OR 666" to be typed. A flowchart of the 15 Puzzle program is shown in Figure 12-15.

Two sample runs of the program are in Figure 12-16. The first number arrangement can be achieved while the second arrangement is one of the 10,461,394,944,000 impossible arrangements.

12.7   A BASIC Prime Number Generating Program

A prime number is an integer that is divisible only by itself and the number 1. One way to determine primes is to use the method developed by Eratosthenes as discussed in section 13.3. This procedure is relatively

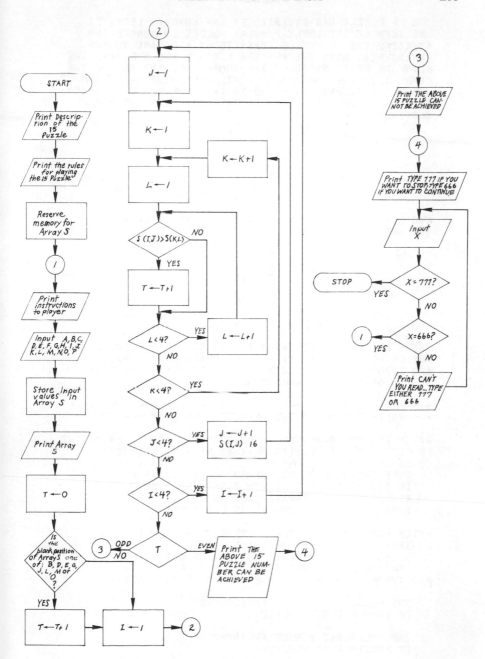

Figure 12-15. Flowchart of the 15 puzzle program.

15 PUZZLE

THE 15 PUZZLE WAS INVENTED BY SAM LOYD IN 1878. IT
HAS BEEN AN EXTREMELY POPULAR PUZZLE IN EUROPE AND
AMERICA. THE 15 PUZZLE CONSISTS OF A SQUARE NUMBER
ARRANGEMENT WITH THE NUMBERS 1 TO 15 AND A BLANK. THE
BLANK IS REPRESENTED BY THE NUMBER 16. ANY ONE OF THE
NUMBERS TO THE IMMEDIATE RIGHT, LEFT, TOP OR BOTTOM OF
THE BLANK SQUARE CAN BE MOVED INTO THE BLANK POSITION.
THE OBJECT OF THE PUZZLE IS TO START WITH THE NUMBER
CONFIGURATION

```
1    2    3    4

5    6    7    8

9    10   11   12

13   14   15   16
```

AND FINISH WITH A DIFFERENT NUMBER ARRANGEMENT SAY
THE ONE SHOWN BELOW

```
4    8    12   14

3    7    11   16

2    6    10   15

1    5    9    13
```

THERE IS ONE SLIGHT CATCH TO THE PUZZLE --- THERE
ARE 10,461,394,944,000 NUMBER ARRANGEMENTS THAT
ARE IMPOSSIBLE TO OBTAIN. THERE ARE ALSO THE SAME
NUMBER OF POSSIBLE ARRANGEMENTS.

--- RULES FOR USING THE 15 PUZZLE PROGRAM ---

AFTER THE PROGRAM TYPES A DESCRIPTION AND RULES OF THE
15 PUZZLE THE FOLLOWING MESSAGE WILL BE PRINTED

    TYPE THE NUMBER ARRANGEMENT TO BE ACHIEVED
    IN ROW ORDER [FIRST ROW FIRST, SECOND ROW
    NEXT, ETC.]. SEPARATE EACH NUMBER BY A
    COMMA. THE BLANK SQUARE IS REPRESENTED BY 16.

AFTER YOU TYPE THE NUMBER ARRANGEMENT TO BE
ACHIEVED AND PRESS RETURN THE PROGRAM WILL
DETERMINE IF THE ARRANGEMENT CAN OR CANNOT BE
ACHIEVED AND WILL PRINT THE NUMBER ARRANGEMENT
AND THE APPROPRIATE OF THE FOLLOWING TWO MESSAGES

    THE FOLLOWING NUMBER ARRANGEMENT
    IS IMPOSSIBLE TO ACHIEVE

    THE FOLLOWING NUMBER ARRANGEMENT
    IS POSSIBLE TO ACHIEVE

YOU MAY INPUT ANOTHER NUMBER ARRANGEMENT BY
TYPING 666 WHEN REQUESTED TO DO SO . YOU MAY
STOP BY TYPING 777 WHEN REQUESTED TO DO SO.

Figure 12-16. Sample runs of 15 puzzle program.

```
TYPE THE NUMBER ARRANGEMENT TO BE ACHIEVED
IN ROW ORDER [ FIRST ROW FIRST, SECOND ROW
NEXT, ETC.]. SEPARATE EACH NUMBER BY A
COMMA. THE BLANK SQUARE IS REPRESENTED BY 16.

? 1,2,3,4,5,6,7,8,9,10,11,12,13,14,16,15,

 1      2      3      4

 5      6      7      8

 9      10     11     12

 13     14     16     15


THE ABOVE 15 PUZZLE NUMBER ARRANGEMENT CAN BE ACHIEVED


TYPE 777 IF YOU WANT TO STOP.
TYPE 666 IF YOU WANT TO CONTINUE.
? 555

CANT YOU READ ---- TYPE EITHER 777 OR 666.
? 666


TYPE THE NUMBER ARRANGEMENT TO BE ACHIEVED
IN ROW ORDER [ FIRST ROW FIRST, SECOND ROW
NEXT, ETC.]. SEPARATE EACH NUMBER BY A
COMMA. THE BLANK SQUARE IS REPRESENTED BY 16.

? 16,15,14,13,12,11,10,9,8,7,6,5,4,3,2,1

 16     15     14     13

 12     11     10     9

 8      7      6      5

 4      3      2      1


THE ABOVE 15 PUZZLE NUMBER ARRANGEMENT CANNOT BE ACHIEVED


TYPE 777 IF YOU WANT TO STOP.
TYPE 666 IF YOU WANT TO CONTINUE.
? 777
```

Figure 12-16. Sample runs of 15 puzzle program (cont'd).

easy to use when one wants to determine only a few prime numbers. However, it would be a rather lengthy operation to determine all the primes less than 200,000 or to determine if 926313 is a prime number.

Figure 12-17 shows a flowchart of a general method for computing primes. The BASIC program for this method is shown in Figure 12-18.

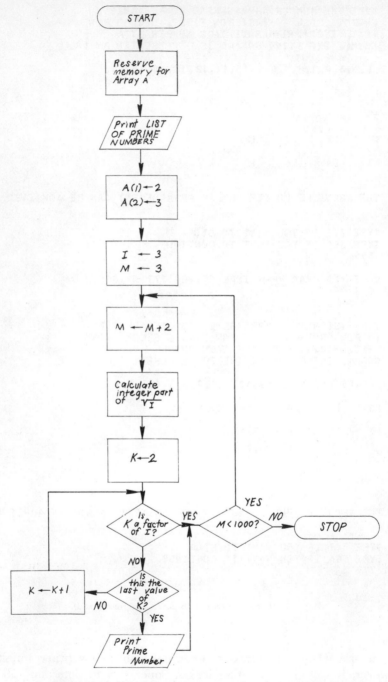

Figure 12-17. Flowchart of the prime number generator program.

```
PRIME      19:16   WSH THUl1/17/66

10 REM PRIME NUMBER GENERATOR (PRIMES < 1000)
20 DIM A(10)
30 PRINT " LIST OF PRIME NUMBERS"
40 PRINT
50 PRINT
54 LET A(1) = 2
56 LET A(2) = 3
60 LET I = 3
70 LET M = 3
80 LET M = M + 2
90 FOR K = 2 TO SQR(M)
100 IF INT(M/K)* K-M = 0 THEN 190
110NEXT K
120LET A(I) = M
130LET I = I + 1
140IF I < 11 THEN 190
150FOR X = 1 TO 10
160PRINT A(X)
170NEXT X
180LET I = 1
190IF M < 1000 THEN 80
200 END
```

Figure 12-18. BASIC program for computing primes.

A few prime numbers that were generated by this program are shown in Figure 12-19.

2	23	59	97	137	179
3	29	61	101	139	181
5	31	67	103	149	191
7	37	71	107	151	193
11	41	73	109	157	197
13	43	79	113	163	199
17	47	83	127	167	211
19	53	89	131	173	223

Figure 12-19. List of prime numbers generated.

12.8   A Tic-Tac-Toe Game to Program

Write a Tic-Tac-Toe program that will first cause playing rules to be printed and then play a game with a human opponent.

The rules of play are as follows:

TIC-TAC-TOE

RULES FOR PLAYING TIC-TAC-TOE WITH A TIME-SHARING COMPUTER

MOVES OF THE HUMAN PLAYER ARE MARKED BY AN X.
THE COMPUTER WILL MARK ITS MOVE WITH AN O.
THE PLAYING BOARD CONSISTS OF A SQUARE DIVIDED INTO

NINE SMALLER SQUARES. THE POSITION OF THE SMALLER
SQUARES ARE LOCATED BY THE NUMBERS 1 THROUGH 9.

1	2	3
4	5	6
7	8	9

THE OBJECT OF THE GAME IS TO GET THREE IDENTICAL
MARKERS IN A STRAIGHT LINE. PLAYS ARE MADE IN ALTER-
NATE MOVES. THE COMPUTER WILL INFORM THE HUMAN
PLAYER WHEN TO MOVE. THE HUMAN PLAYER CAN ALWAYS
MAKE THE FIRST MOVE.

THE HUMAN PLAYER INFORMS THE COMPUTER OF A MOVE
BY TYPING THE POSITION NUMBER AND PRESSING THE RE-
TURN BUTTON. THE COMPUTER WILL TYPE THE BOARD
CONFIGURATION AFTER EACH PLAY AND WILL INFORM THE
HUMAN PLAYER WHO WINS THE GAME.

A sample game of play is used to illustrate the messages that the program must type and the format of the Tic-Tac-Toe board.

START OF TIC-TAC-TOE GAME
HUMANS MOVE IS? 5
—　　—　　O

—　　X　　—

—　　—　　—

HUMANS MOVE IS? 4
—　　—　　O

X　　X　　O

—　　—　　—

HUMANS MOVE IS? 4
HOW ABOUT PICKING A POSITION THAT IS NOT ALREADY
OCCUPIED:
HUMANS MOVE IS? 9
O　　—　　O

X　　X　　O

—　　—　　X

HUMANS MOVE IS? 8
O　　O　　O

X　　X　　O

—　　X　　X

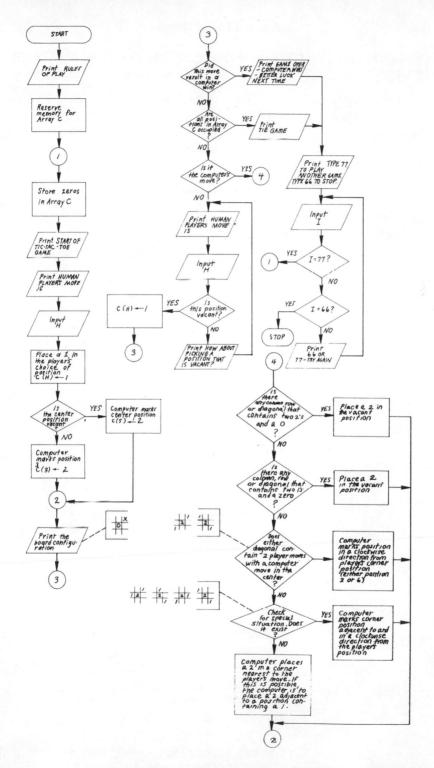

Figure 12-20. Flowchart of the Tic-Tac-Toe program.

GAME OVER—COMPUTER WINS
BETTER LUCK NEXT TIME

A flowchart of the Tic-Tac-Toe program is shown in Figure 12-20.

A close observation of this flowchart will indicate that it does not contain a *human wins* printout. This flowchart procedure allows the human player to *tie* but not to beat the computer.

12.9   A Knight's Tour to Program

The rules of play in Figure 12-21 describe the fascinating tour of the knight on a chessboard.

```
THE STRANGE MOVE OF THE KNIGHT MAKE HIS
OPERATIONS FASCINATING. HE IS PERMITTED TO
MOVE TWO OR ONE ROWS UP OR DOWN AND
ONE OR TWO COLUMNS LEFT OR RIGHT AS
ILLUSTRATED BELOW

    -    -    -    -    -    -    -    -

    -    -    -    -    -    -    -    -

    -    -    -    -    -    -    -    -

    -    X    -    X    -    -    -    -

    X    -    -    -    X    -    -    -

    -    -    K    -    -    -    -    -

    X    -    -    -    X    -    -    -

    -    X    -    X    -    -    -    -

A KNIGHTS TOUR IS WHERE THE KNIGHT
TRAVELS TO EVERY SQUARE ON THE
CHESSBOARD WITHOUT BEING ON ANY
SQUARE TWICE.

FOR EXAMPLE , ONE SUCH TOUR IS SHOWN BELOW

    1   42   13   26    3   60   15   28

   24   37    2   41   14   27    4   61

   43   12   25   38   59   62   29   16

   36   23   40   63   48   51   56    5

   11   44   49   52   39   58   17   30

   22   35   64   47   50   55    6   57

   45   10   33   20   53    8   31   18

   34   21   46    9   32   19   54    7

THE PATH OF THIS KNIGHTS TOUR STARTED IN
CELL[1,1] AND ENDED IN CELL[6,3].
```

Figure 12-21. Rules of play for a knight's tour.

Write a Knight's Tour program using the rules-of-play printout and the flowchart shown in Figure 12-22.

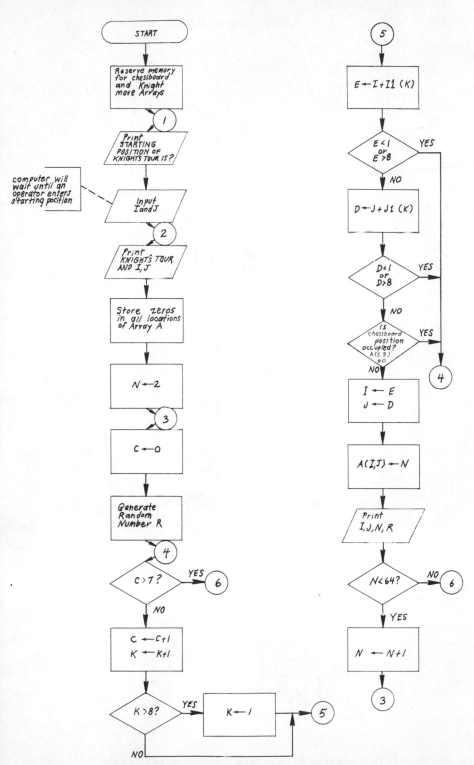

Figure 12-22. Flowchart of the knight's tour program.

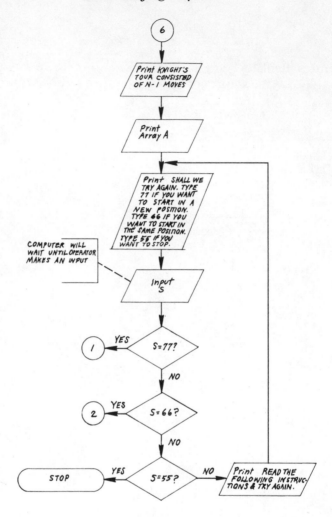

Array A is used to represent the checkerboard
Array I1 and J1 represent the moves of the Knight

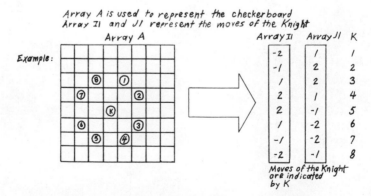

Moves of the Knight
are indicated
by K

Figure 12-22. Flowchart of the knight's tour program (cont'd).

12.10   A Roulette Game to Program

Write a Roulette program that will cause the following rules of play to be typed. A flowchart of this program is shown in Figure 12-23.

ROULETTE BETS AND ODDS

EVEN MONEY ON RED, BLACK, EVEN, ODD, 1 TO 18, OR
19 TO 36
2 TO 1 ON 1ST, 2ND OR 3RD TWELVE
2 TO 1 ON ANY COLUMN
5 TO 1 ON ANY 6 NUMBERS
6 TO 1 ON 0, 00, 1, 2 AND 3
8 TO 1 ON ANY 4 NUMBERS
11 TO 1 ON ANY 3 NUMBERS
17 TO 1 ON ANY 2 NUMBERS
35 TO 1 ON ANY SINGLE NUMBER

LETS PLAY ROULETTE

BET (IN DOLLARS) WHEN REQUESTED TO DO SO.
TYPE A ZERO IF YOU DO NOT WANT TO BET ON
ANY CHOICE. 00 IS REPRESENTED BY 37.

The program will cause the following messages to be typed. Each message is to be answered by the human operator typing either 0 or the amount in dollars that he wants to bet. The number 37 is used by the program to represent the roulette number 00.

IF YOU WANT TO BET ON RED—TYPE THE AMOUNT
IF YOU WANT TO BET ON BLACK—TYPE THE AMOUNT
IF YOU WANT TO BET ON EVEN—TYPE THE AMOUNT
IF YOU WANT TO BET ON A LOW NUMBER (1-18)—
   TYPE THE AMOUNT
IF YOU WANT TO BET ON A HIGH NUMBER (19-36)—
   TYPE THE AMOUNT
IF YOU WANT TO BET ON A TWELVE—TYPE 1, 2 OR 3
   AND THE AMOUNT SEPARATED BY A COMMA
IF YOU WANT TO BET ON A COLUMN—TYPE 1, 2 OR 3
   AND THE AMOUNT SEPARATED BY A COMMA
IF YOU WANT TO BET ON A SINGLE NUMBER—TYPE THE
   NUMBER AND THE AMOUNT
NO MORE BETS—THE WHEEL IS SPINNING.
THE BALL STOPPED AT __.

The program is to type one of the following messages depending on whether the player won or lost:

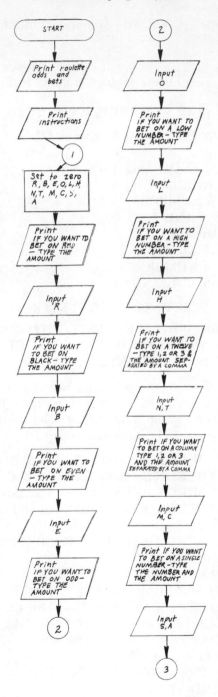

Figure 12-23. Flowchart of the roulette program.

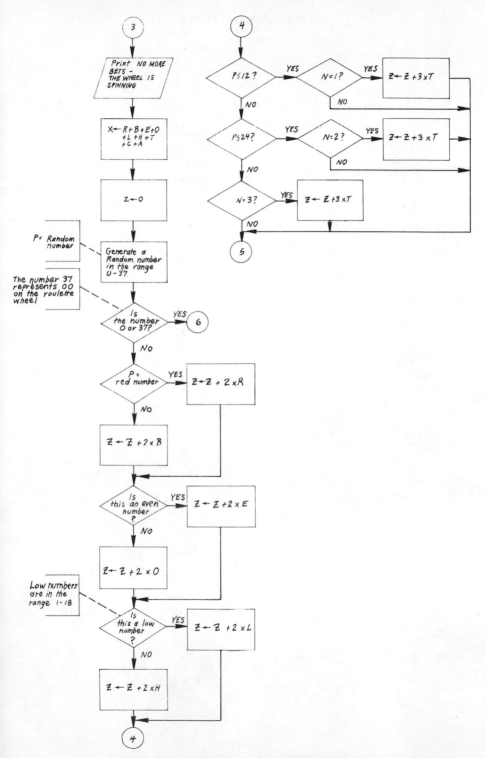

Figure 12-23. Flowchart of the roulette program (cont'd).

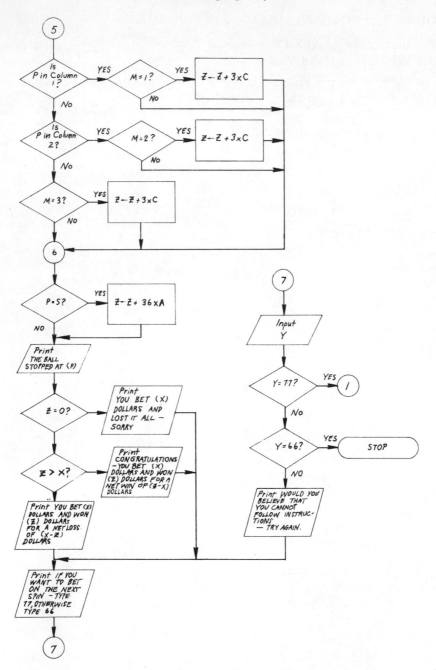

Figure 12-23. Flowchart of the roulette program (cont'd).

YOU BET __ DOLLARS AND LOST IT ALL—SORRY.

YOU BET __ DOLLARS AND WON __ DOLLARS FOR A
NET WIN OF __ DOLLARS.

YOU BET __ DOLLARS AND WON __ DOLLARS FOR A NET
LOSS OF __ DOLLARS.

12.11 An Order 4 Magic Square Generator to Program

Write a program that will cause the following rules of play to be typed. A flowchart of the program is shown in Fig. 12-24.

ORDER 4 MAGIC SQUARE

THIS PROGRAM WILL GENERATE A 4 BY 4 MAGIC SQUARE BY USING BERGHOLTS GENERAL FORM. THE ALPHABETICAL VALUES IN THE SQUARE (A, B, C, D, W, X, Y, Z) MAY BE REPLACED BY ANY POSITIVE OR NEGATIVE NUMERICAL VALUE.

A − W	C + W + Y	B + X − Y	D − X
D + W − Z	B	C	A − W + Z
C − X + Z	A	D	B + X − Z
B + X	D − W − Y	A − X + Y	C + W

THE ABOVE SQUARE WILL ALSO PRODUCE SYMMETRICAL AND PANDIAGONAL SQUARES. THE SQUARE WILL BE SYMMETRICAL IF

$$A + D = B + D$$
$$W + Y = X − Y = Z$$

THE SQUARE WILL BE PANDIAGONAL IF

$$\frac{A − B − C + D}{2} = Z − Y = W = X$$

YOU MAY GENERATE A SQUARE BY TYPING THE TYPE AND VALUES FOR A, B, C, D, W, X, Y, Z WHEN REQUESTED TO DO SO.

TYPE 1 FOR A PANDIAGONAL SQUARE
TYPE 2 FOR A SYMMETRICAL SQUARE
ANY OTHER VALUE FOR TYPE WILL CAUSE
THE PROGRAM TO GENERATE A 4 BY 4
MAGIC SQUARE.

THE PROGRAM CHECKS THE RELATIONSHIPS FOR SYMMETRI-

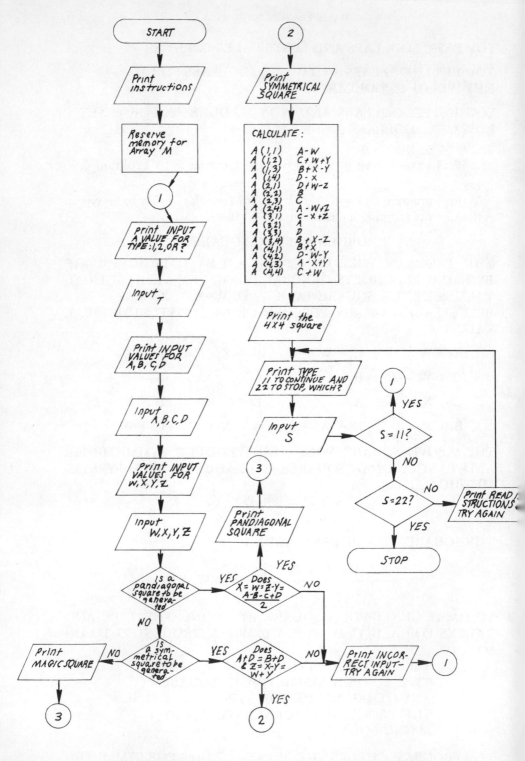

Figure 12-24. Rules of play for a 4 by 4 magic square.

CAL AND PANDIAGONAL SQUARES AND WILL PRINT A MESSAGE IF THE VALUES OF A, B, C, D, W, X, Y, Z DO NOT ALLOW THE RELATIONSHIPS TO EXIST.

After typing the playing rules, the program types three messages:

INPUT A VALUE FOR TYPE: 1, 2, OR?
INPUT VALUES FOR A, B, C, D?
INPUT VALUES FOR W, X, Y, Z?

After the human operator supplies the values for the previous messages, the program will compute and type the magic square. After typing the square the message

TYPE 11 TO CONTINUE AND 22 TO STOP. WHICH?

is typed. The program will stop if the human operator types a 22 and start again if 11 is typed. If the operator types a value other than 11 or 22, the program will type

READ INSTRUCTIONS AND TRY AGAIN

and give the operator another chance to type a correct value.

CHAPTER 13

Betting Systems

THE READER MAY WISH to incorporate some method of betting into the gambling games described in Chapters 6 and 10. The following betting systems may be used with most games of chance, including Roulette, Craps, Faro, Boule, Chinese Fan Tan, Chemin de Fer, Baccarat, Trente et Quarante and Chuck-A-Luck.

Gamblers are an unpredictable lot of people and have been known to tear up cards, swallow dice, wreck the furniture, blow on dice and, during the early development of the American West, shoot the winner of a game. Other gamblers carry rabbit foot charms, or other tokens in an attempt to bribe *lady luck.*

The average gambler in Nevada gambling houses places his bet and *hopes for the best.* The professional gambler attempts to use a betting scheme instead of depending upon some supernatural force. In games in which skill plays a large part, the gambler can minimize his losses by increasing his gaming skill. However, in games of pure chance, such as Roulette, Faro or Craps, the gambler may try to work out a scientific *system* in an attempt to increase his chance of winning.

Gambling houses usually welcome *system* players, since no matter what system is being used, the house will eventually win everything that is bet. A few systems have proved to be reliable for a short period of time. A system based on the mechanical imperfections of a Monte Carlo roulette wheel resulted in a player winning almost $200,000 before the casino realized that the wheel was defective. This system ended when the casino replaced the wheel with one that was perfectly balanced. Today, in most casinos, roulette wheels go through a periodic maintenance check every day, thus *practically* eliminating a system of this type. Another system which was given world wide publicity in 1964 was Edward Thorp's Blackjack system, described in Chapter 1.

Many systems seem to work in theory but fail when put to test on the gaming table. Other systems fail because the player does not always follow his original system. The excitement of many gambling games makes the player throw away his original system and play some hunch. It really doesn't make much difference because the "house" always wins in the long run.

A table of roulette values is presented as a tool for the reader to use in understanding the various betting systems. The roulette table shows the winning numbers, in the order that they appeared, of 100 turns of a double zero roulette wheel. This table of 100 sample plays also specifies whether the winning numbers were either red or black, odd or even, high or low, in the 1st, 2nd or 3rd dozen, in the 1st, 2nd or 3rd column or that they were 0 or 00. Roulette is the system player's game. It is for this reason that it is used in the descriptions of the various betting systems. In fact, system play at the Monte Carlo roulette wheels is so popular that the casino publishes a *Green Book* describing the day-by-day play of their roulette wheels. Of course this book only encourages system players to become more active in roulette play, and as everyone knows, system players will eventually lose all money bet, like the non-system player, although they may hang on to their money a little longer.

ROULETTE TABLE

PLAYS 1 through 25

PLAY	NUMBER	COLOR	ODD-EVEN	HIGH-LOW	DOZEN	COLUMN
1st	10	B	E	L	1	1
2nd	22	B	E	H	2	1
3rd	34	R	E	H	3	1
4th	11	B	O	L	1	2
5th	2	B	E	L	1	2
6th	27	B	O	H	3	3
7th	19	R	O	H	2	1
8th	28	B	E	H	3	1
9th	30	R	E	H	3	3
10th	33	B	O	H	3	3
11th	9	R	O	L	1	3
12th	18	R	E	L	2	3
13th	8	B	E	L	1	2
14th	13	B	O	L	2	1
15th	2	B	E	L	1	2
16th	1	R	O	L	1	1
17th	23	R	O	H	2	2
18th	3	R	O	L	1	3
19th	35	B	O	H	3	2
20th	6	B	E	L	1	3
21st	8	B	E	L	1	2
22nd	13	B	O	L	2	1
23rd	24	B	E	H	2	3
24th	4	B	E	L	1	1
25th	21	R	O	H	2	3

PLAYS 26 through 50

PLAY	NUMBER	COLOR	ODD-EVEN	HIGH-LOW	DOZEN	COLUMN
26th	31	B	O	H	3	1
27th	1	R	O	L	1	1
28th	10	B	E	L	1	1
29th	0	G	—	—	—	—
30th	15	B	O	L	2	3
31st	16	R	E	L	2	1
32nd	2	B	E	L	1	2

PLAY	NUMBER	COLOR	ODD-EVEN	HIGH-LOW	DOZEN	COLUMN
33rd	4	B	E	L	1	1
34th	33	B	O	H	3	3
35th	3	R	O	L	1	3
36th	27	R	O	H	3	3
37th	15	B	O	L	2	3
38th	29	B	O	H	3	2
39th	14	R	E	L	2	2
40th	25	R	O	H	2	1
41st	10	B	E	L	1	1
42nd	27	R	O	H	3	3
43rd	18	R	E	L	2	3
44th	35	B	O	H	3	2
45th	22	B	E	H	2	1
46th	6	B	E	L	1	3
47th	34	R	E	H	3	1
48th	12	R	E	L	1	3
49th	36	R	E	H	3	3
50th	19	R	O	H	2	1

PLAYS 51 through 75

PLAY	NUMBER	COLOR	ODD-EVEN	HIGH-LOW	DOZEN	COLUMN
51st	9	R	O	L	1	3
52nd	20	B	E	H	2	2
53rd	24	B	E	H	2	3
54th	9	R	O	L	1	3
55th	6	B	E	L	1	3
56th	7	R	O	L	1	1
57th	27	R	O	H	3	3
58th	15	B	O	L	2	3
59th	33	B	O	H	3	3
60th	7	R	O	L	1	1
61st	20	B	E	H	2	2
62nd	24	B	E	H	2	3
63rd	2	B	E	L	1	2
64th	19	R	O	H	2	1
65th	36	R	E	H	3	1
66th	16	R	E	L	2	1
67th	27	R	O	H	3	1
68th	19	R	O	H	2	1
69th	35	B	O	H	3	2
70th	30	R	E	H	3	3
71st	13	B	O	L	2	1
72nd	32	R	E	H	3	2
73rd	2	B	E	L	1	2
74th	14	R	E	L	1	2
75th	27	R	O	H	3	3

PLAYS 76 through 100

PLAY	NUMBER	COLOR	ODD-EVEN	HIGH-LOW	DOZEN	COLUMN
76th	21	R	O	H	2	3
77th	27	R	O	H	3	3
78th	30	R	E	H	3	3
79th	32	R	E	H	3	2
80th	33	B	O	H	3	3
81st	26	B	E	H	3	2
82nd	31	B	O	H	3	1
83rd	00	G	—	—	—	—
84th	16	R	E	L	2	1
85th	31	B	O	H	3	1
86th	11	B	O	L	1	2
87th	26	B	E	H	3	2
88th	21	R	O	H	2	3
89th	7	R	O	L	1	1
90th	31	B	O	L	3	3
91st	3	R	O	L	1	2
92nd	17	B	O	L	2	1
93rd	4	B	E	L	1	3
94th	15	B	O	H	2	1
95th	25	R	O	H	3	3
96th	9	R	O	L	1	3
97th	29	B	O	H	3	2
98th	4	B	E	L	1	1
99th	19	R	O	H	2	1
100th	22	B	E	H	2	1

The reader should realize that the following systems apply to games other than Roulette. In fact they may be used with Craps, Faro, Boule or most other games that have events with even or near-even chance of occurring.

13.1 Doubling Up System

A common method of betting on even-money chances is to double the preceding bet if you lose it. This system is based on the theory that if it has an even chance of occurring, like the black color at Roulette, it must come up black eventually, and the longer it is in coming, the more certain it is to come; therefore, the more you may bet upon it.

Consider the following betting procedure. Say your first bet is for $1 and that you lost. You would then double your bet and now wager $2 to even things up. Let's say you lose again. You now must bet $4. If you lose this bet then the next bet must be $8. Say you lose this bet and wager $16 for the next bet. This doubling would continue as long as you are losing or until the *house limit* was reached. The house limit on most single bets on the roulette wheel is around $500. It doesn't take too long a losing streak to reach the house limit. A powers of 2 table mav be used to express the money lost in the *Doubling Up System.*

Play	Player Bets and Loses	Cumulative Losses of Player
1	1	$ 1
2	2	3
3	4	7
4	8	15
5	16	31
6	32	63
7	64	127
8	128	255
9	256	511
10	512	1023
11	1024	2047
12	2048	4095
13	4096	8191
14	8192	16383
15	16384	32767

When using the Doubling Up System, the original bet must be very small, because, as previously stated, the house limit is reached after only a few losses. Several long runs of one color have been recorded. The author has seen Black appear 22 times before a Red number finally showed. One source claims that Red came up 23 times and the 24th number was 0. Another source reports that a 33 color run appeared. Long runs of this type are not too common but the system player must not forget that they do occur.

13.2 The Simple Martingale

The *Simple Martingale* is the basic Doubling Up System previously described. For example, if a player was betting on black after red had appeared in the past four successive turns of the wheel, and red appears again, thus making a string of five successive red appearances, then the player would double the bet and place it on black. If this bet loses, then the player must double the bet again and place this amount on black. However, if the last bet had won, the player would start afresh with a bet of one.

13.3   The Great Martingale

The *Great Martingale* is a system where the bet is doubled and one added after every losing play. Thus a succession of losing plays would result in the following plays:

Play	Player Bets	Cumulative Losses of Player
1	1	1
2	$(2 + 1) = \quad 3$	4
3	$(6 + 1) = \quad 7$	11
4	$(14 + 1) = \quad 15$	26
5	$(30 + 1) = \quad 31$	57
6	$(62 + 1) = \quad 63$	120
7	$(126 + 1) = \quad 127$	247
8	$(254 + 1) = \quad 255$	502
9	$(510 + 1) = \quad 511$	1013
10	$(1022 + 1) = 1023$	2036
11	$(2046 + 1) = 2047$	4083
12	$4094 + 1) = 4095$	8178

The Great Martingale will reach the *house limit* one play sooner than the Simple Martingale would. The main advantage of this system is to provide a rapid double-up.

13.4   The Reverse Martingale

The *Reverse Martingale* is a system where bets are doubled up after a loss, as in other martingale systems, but after each successive win, the winnings are left to ride for the next bet. This procedure would continue until *eight* successive wins occur, at which time the winnings are removed and the player would start afresh with a new bet of one unit.

13.5   The D'Alembert System

A player using the *D'Alember System* would start his first bet with a certain number of chips, say 10. Every time he wins a bet, he decreases the bet by one chip, making it nine chips. Every time he loses, he increases the bet by one chip, going to 11. If he wins a second time, he goes back to nine chips again. If he wins 10 bets and loses 10 bets, no matter in what order the events happen, he will be 10 chips ahead, provided the game does not run against him for a time. In this case he would be continually betting larger and larger amounts, and would have to continue betting until the tide turned, which it might not do for months.

13.6   The Biarritz System

The *Biarritz System* is strictly for roulette players who like to bet on single numbers where the payoff is 35 to 1. The system is played by watching at least 114 consecutive spins of the wheel before placing the first bet. The 114 figure is determined by multiplying the number of numbers on the wheel by three. The system is based on the idea that 114 spins is a sufficient number of spins to allow every number to appear at least once. The player determines which number to eventually place a bet on by logging the 114 numbers and betting on the number that appeared the least number of times. The player is counting on this number appearing again during the next 34 spins of the wheel.

The player may make 34 one-unit bets before losing. If the number came up on the 34th spin, he would win at 35 to 1 odds, thus winning 1 unit more than originally bet. However, if the number appeared before the 34th spin, he would profit even more. If the number did not appear the player loses.

13.7   The Cuban System

Another method used on the roulette wheel is the *Cuban System*. This system is based on the fact that the third column on the roulette layout has four black numbers and eight red numbers, as shown in Figure 13-1.

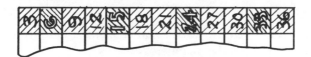

Figure 13-1. Third column on the roulette layout.

A player places an even money bet on BLACK and a 2 to 1 bet on Column 3. The reasoning behind these bets is that they will produce more than 76 units (which is the amount bet) in every 38 spins of the wheel.

13.8   The Labouchere System

The *Labouchere System* uses the following procedure. Start with any number of chips, any five, and write them down in a vertical column.

$$1$$
$$2$$
$$3$$
$$4$$
$$5$$

The player bets an amount equal to the sum of the numbers on the top and bottom of the column. After a win, cross out the numbers used in the bet. After a loss, write the amount of the bet at the bottom of the column. When all numbers are crossed out, start with a new list.

For example, using a list of five numbers a player would make a first bet of six chips. Assume that the player wins, thus resulting in the top and bottom numbers being crossed out.

$$\cancel{1}$$
$$2$$
$$3$$
$$4$$
$$\cancel{5}$$

The next bet would be the sum of 2 and 4, thus 6 chips would be bet. Assume that the player lost this bet. The number 6 would be added to the bottom of the list.

$$\cancel{1}$$
$$2$$
$$3$$
$$4$$
$$\cancel{5}$$
$$6$$

The next bet would be 8 chips, the sum of 2 and 6. All future bets would be determined in a similar manner.

13.9   The Ascot System

The *Ascot System* uses a progression of 3, 6, 9, 12, 20, 25, 35, 50, 75, 125, 200.

The player starts with a bet of 25 chips and goes down the scale whenever he loses, up the scale as long as he wins. The player would start afresh with a bet of 25 immediately following the 200 chip bet.

CHAPTER 14

Random Numbers

MANY OF THE GAMES described in this book require the use of random numbers: Keno, Roulette, Dice, Bingo, Boule, Chuck-A-Luck, Wheel of Fortune, Crown and Anchor, Baccarat, Chemin de Fer and others. Several methods of generating these numbers are covered in this chapter.

Random numbers occur in many games. A roulette wheel is an excellent random number generator that will produce numbers between 1 and 38 (assuming that 0 and 00 represented the numbers 37 and 38 respectively). The toss of a coin is a random number generator with an output of the numbers 1 and 2 (assuming that *heads* stands for 1 and tails stands for 2). The roll of a cubical die will produce random numbers in the range 1 through 6. A Keno cage will produce random numbers in the range 1 through 80. Even a well-shuffled deck of cards may be used to produce random numbers in the range 1 through 52 (assuming that each card represented a unique number in the range 1-52).

Many simulation and game-playing computer programs use random numbers. A common source for obtaining these numbers is from a table such as the Rand Table.* A table look-up scheme may be useful in some game programs where only a few numbers are needed. However, if many numbers are needed, it is advisable to produce the random numbers by a computational method. A long table of random numbers in computer storage is uneconomical.

The digital computer is a good random number generator. There are many computational schemes that produce these numbers. The numbers generated by computers in a deterministic manner are termed *pseudo-*

---

* RAND Corporation. *One Million Random Digits and 1000,000 Normal Deviates.*

random numbers. However, these pseudo-random numbers have been subjected to many statistical tests of randomness and for practical purposes may be considered random numbers. In this book, the term *random number* refers to *pseudo-random number*, and the term *random number generator* refers to a method of generating *pseudo-random numbers*.

A survey of several generating methods follows.

14.1   Survey of Random Number Generating Methods

Mid Square Technique. This technique was first introduced by Von Neumann. It consists of taking an *n* digit number, squaring it, and using the middle *n* digits as the random number. This number is in turn squared and the process repeated to determine the next number, and so forth.

The selection of the starting number is extremely important when using this method as the process has a tendency to generate zeroes if started with certain numbers, i.e. 165.

An example of how random numbers can be generated by this method follows.

Take the number 76 and obtain the square.

$$76^2 = 5776$$

Use the middle two digits as the computed random number. Then square 77.

$$77^2 = 6929$$

which gives us a random number of 92. Continuing in this manner will result in the generation of the following random numbers:

$$76,77,92,46,11,12,14,19,36,29,84,07, \ldots$$

Table Look-Up. A table look-up scheme is useful when only a small number of random numbers are required. A table consisting of a long sequence of the digits 0 to 9 can be arranged so that the probability exists that every nth entry is statistically independent of the other entries. The program would simply reference the table and select the random numbers in order.

Multiplicative Generator. The multiplicative process for generating random numbers consists of multiplying a constant C by a number R_n. This method generates a random number R_{n+1}.

$$R_{n+1} = C \cdot R_n \ (\text{mod} \ x)$$

where mod x represents the word length of the computer or the limit of the desired sequence and the constant C is usually odd and relatively prime to the base. This constant will determine the period of the generated numbers. A suitable choice for C, 1) on binary computers is an odd power of 5, say 5^7 or 5^{13}, and 2) on decimal computers is an odd power of 7, say 7^{13} or 7^{17}.

Lehmer's Method. The following program was written for the Univac-1206 computer. The method multiplies the number 12345678 by 23, and subtracts the top two digits of the product from the remaining digits of the product. The resulting random number is located in RN.

RANDOM	ENTRY		
	ENT	Q*W(RN)	STARTING CONSTANT
	MUL	W(RNM)	CONSTANT MULTIPLIER
	ENT	Q*A	
	ENT	LP*W(RN1)	PICK-OFF UPPER 2 DIGITS
	RSH	A*17D	
	STR	A*W(RN)	
	LSH	Q*7	PUT DIGITS IN PROPER POSITION
	SUB	Q*W(RN)	SUB DIGITS FROM REST OF PRODUCT
	STR	Q*W(RN)	NEW RANDOM NUMBER
	EXIT		
RN	0374676140		.12345678D SCALED 29
RNM	0727024340		.23D SCALED 29
RN1	1760000000		MASKING CONSTANT

Mixed Congruence Method. A variation of the multiplicative method is given by

$$R_{n+1} = C \cdot R_n + K \ (\mathrm{mod}\,x)$$

where mod x represents the word length of the computer or the limit of the desired sequence, C is a constant of the form $2^n + 1$, K is a positive prime number, R_n is the current random number and R_{n+1} is the random number being calculated. The following example is used to illustrate this method.

Assume that

$$C = 32769$$
$$K = \quad 37$$
$$\mathrm{Mod}\,x = 16777216$$

The resultant model is

$$R_{n+1} = 32769R_n + 37 \ (\mathrm{mod}\ 16777216)$$

Let $R_n = 1212490$, then the first few numbers generated are

$$03637359$$
$$07274644$$
$$12124345$$

A flowchart of this method is shown in Figure 14-1.

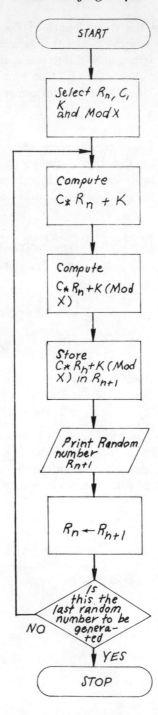

Figure 14-1. Flowchart of the mixed Congruence method.

Fibonacci Sequence. This method will generate random numbers *faster* than the multiplication generator and it has a large period. The method is defined as

$$R_{n+2} = R_{n+1} + R_n \ (\mathrm{mod}\ x)$$

where mod x represents the word length of the computer, R_{n+2} is the random number being calculated, R_{n+1} is the last calculated random number, and R_n is the random number that was calculated prior to R_{n+1}.

CHAPTER 15

Game Playing Exercises

1. What is an algorithm? Name 4 game programs that have known algorithms.
2. Define heuristic. Name three game programs that are usually programmed by using a heuristic process.
3. Write an algorithm for finding the average of a set of numbers.
4. Construct a set of pentominoes from cardboard; attempt to play this game. You may wish you had the help of a computer long before you come up with a solution.
5. Starting with a knight in the position indicated in Figure 15-1, complete a tour of the chess board.
6. A popular ancient puzzle is called the *Tower of Honoi*. The puzzle consists of a horizontal board with three vertical pegs, as shown in Figure 15-2.

   Several discs of different sizes are arranged on one of the pegs. The largest disc is at the bottom and the smallest is at the top of the peg. The object of the puzzle is to transfer all of the discs

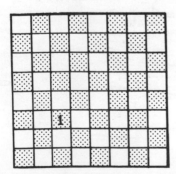

Figure 15-1. Starting position for a knight's tour.

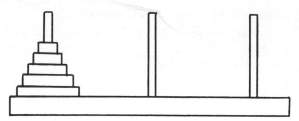

Figure 15-2. Tower of Honoi.

from the first peg to one of the others in such a way that the final arrangement is identical to the original one. Only *one* disc can be moved at one time and no disc can ever rest on one smaller than itself. A game with 64 discs would require $2^{64}-1$ transfers.

Can you see how a computer may be used to solve a solution to this puzzle? One should use a small number of discs, to start, say five, as it would be easier to check the solution. Five discs can be transferred from one peg to another in 2^5-1 or 31 moves. It takes 63 moves to move 6 discs, 127 moves to move 7 discs and 255 to move 8 discs.

7. Purchase a 15 Puzzle at a local department or drug store and see if it is possible to obtain the number configurations in Figure 15-3.

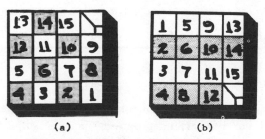

Figure 15-3. 15 Puzzle number configurations.

8. Do you think a computer could be used to solve a recreational logic problem such as the following?

Census taker. The census taker, while taking head counts in a nearby village, questioned the one-armed owner of a small rundown shack. He pointed to another one-armed gent who was asleep. "Who is he?", he asked. The one-armed man replied, *"Brothers and sisters have I none, but that man's father is my father's son."*

The object of the puzzle is to determine who the sleeping man is. *(Note: the sleeping man was his son.)*

9. Play the game of Nim using the following coins.

00
000
0000

Can you discover the winning first move?

10. What is the estimated number for the possible moves in an average game of Checkers?

11. What is the estimated number for the possible moves in an average game of Chess?

12. Name ten games that can be played with a digital computer.

13. Determine which one of the number arrangements in Figure 15-4 represents a magic square.

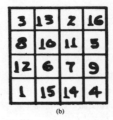

(a)   (b)

Figure 15-4. Number arrangements.

14. Write a 15 Puzzle program in a programming language other than FORTRAN.

15. Design an input data card for the 34 Puzzle checking program that will cause the program to check the following number arrangement:

6	4	14	7
9	15	1	12
3	10	8	13
16	5	11	2

The above number arrangement does not meet the requirements of the 34 Puzzle checking program since the sum of Row 1, Row 2, Diagonal 1 and 4-cell group 2,1 does not equal 34.

16. Make a set of PICK-A-NUMBER cards for detecting any number from 1 through 63.

17. Design a set of PICK-A-NUMBER cards that may be used to select any number between 1 and 100.

18. Rewrite the 3 Coin Game program in a different programming language.

19. Write a FORTRAN II program that will generate prime numbers from the polynomial $x^2 + x + 41$ when x varies from 0 through 40.

20. Modify the Sieve of Eratosthenes program to allow for the first 1000 numbers.

21. The polynomial $x^2 + x + 17$ may be used to generate 16 prime numbers when x = 0, 1, 2, 3, 4, 5, . . . , 14, 15. Write a FORTRAN II program that will do this.

22. Using the rules outlined in Table 5-2, determine if the following number arrangement is a valid solution to the 15 Puzzle:

1	6	3	7
4	2	9	11
12	14	5	10
13	8	15	

23. Determine if the number arrangement in Figure 15-5 is a Magic Square.

15	16	22	3	9
8	14	20	21	2
1	7	13	19	25
24	5	6	12	18
12	23	4	10	11

Figure 15-5. Number arrangement.

24. Write a Blackjack program in a programming language other than FORTRAN.

25. What is the magic number of a 4 by 4 Magic Square? Assume that the numbers begin with 18.

26. Write a FORTRAN statement that will compute the magic number of a specified square.

27. The following diagram illustrates a form that may be used to generate one of the 8 possible number arrangements of a 3 by 3 Magic Square.

$a + b$	$a - (b + c)$	$a + c$
$a - (b - c)$	a	$a + (b - c)$
$a - c$	$a + (b + c)$	$a - b$

Hand-calculate a Magic Square by using the values: $a = 5$, $b = 1$, and $c = 3$.

28. Hand-calculate a 9 by 9 Magic Square using the De la Loubere generating procedure.

29. Hand-calculate a 3 by 3 Magic Square using the method of De la Loubere.

30. Rewrite the De la Loubere Magic Square generating program in FORTRAN IV to compute Odd Order Magic Squares up to size 31 by 31.

31. Construct a 3 by 3 Magic Square from the numbers of the geometrical progression

1, 2, 4, 8, 16, 32, 64, 128, 156

so that the *product* of each row, column and main diagonal equals 4096.

32. Rewrite the Agrippa Magic Square generating program in FOR-TRAN IV to compute magic squares up to size 31 by 31.

33. Construct a magic square by placing a *different number* in each vacant cell of the square in Figure 15-6 so that each column, row and main diagonal adds up to the magic number 15.

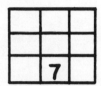

Figure 15-6. Making a magic square.

34. Write a FORTRAN IV Magic Square generating program to compute a 25 by 25 Magic Square starting with the number 38.

35. Form a 3 by 3 Magic Square from the following powers of 2:

$$2^0, 2^1, 2^2, 2^3, 2^4, 2^5, 2^6, 2^7, 2^8$$

so that the *product* of each row, column and main diagonal equals 2^{12}. Note the similiarity between this exercise and Exercise 31.

36. Write a FORTRAN IV program that will interchange rows 3 and 7 of the 9 by 9 Magic Square in Figure 15-7.

37	48	59	70	81	2	13	24	35
36	38	49	60	71	73	3	14	25
26	28	39	50	61	72	74	4	15
16	27	29	40	51	62	64	75	5
6	17	19	30	41	52	63	65	76
77	7	18	20	31	42	53	55	66
67	78	8	10	21	32	43	54	56
57	68	79	9	11	22	33	44	46
47	58	69	80	1	12	23	34	45

Figure 15-7. 9 by 9 magic square.

37. Fill in the blank square with numbers determined by the relationships in Figure 15-8. The square will be magic if the numbers are correct.

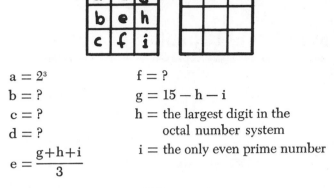

$a = 2^3$

$b = ?$

$c = ?$

$d = ?$

$e = \dfrac{g+h+i}{3}$

$f = ?$

$g = 15 - h - i$

$h =$ the largest digit in the octal number system

$i =$ the only even prime number

Figure 15-8. Making a magic square.

38. Write a FORTRAN IV program that will generate all odd and even magic squares up to size 12 by 12. (Hint—use a main program that calls on various subroutine generating subprograms. One subroutine could generate odd order squares: 3 by 3, 5 by 5, 7 by 7, 9 by 9 and 11 by 11. Another subroutine could generate even squares of order 4, 8 and 12 and the remaining squares of order 6 and 10 could be computed by a third subroutine.)

39. A Magic Square will result if the blank square is filled in with the correct numbers resulting from the relationships in Figure 15-9.

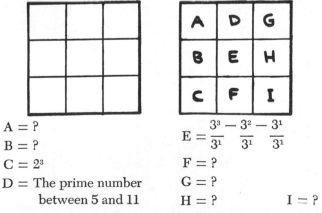

$A = ?$

$B = ?$

$C = 2^3$

$D =$ The prime number between 5 and 11

$E = \dfrac{3^3}{3^1} - \dfrac{3^2}{3^1} - \dfrac{3^1}{3^1}$

$F = ?$

$G = ?$

$H = ?$ $I = ?$

Figure 15-9. Making a magic square.

40. Write a FORTRAN IV program that will generate Domino Magic Squares.

Answers to Odd Number Exercises

1. Algorithm—is a list of instructions that specifies a sequence of operations which will give a correct answer to a problem of a given type.
3. (1) Add the given numbers.
   (2) Count the given numbers.
   (3) Divide the answers to (1) by the answer to (2).
5. Knight's Tour of the Chessboard.

64	23	52	39	62	5	50	37
41	28	63	24	51	38	61	4
22	53	40	27	6	3	36	49
29	42	25	2	17	14	9	60
54	21	16	13	26	7	48	35
43	30	1	18	15	10	59	8
20	55	32	45	12	57	34	47
31	44	19	56	33	46	11	68

7. a is possible and b is impossible.

9. The only way that the first player can be assured of winning is by leaving one coin in the bottom row after his first move.

11. 10^{120}

13. Number arrangement *a* represents a magic square.

15.

17.

	CARD 1						CARD 2		
1	21	41	61	81		2	22	42	62	82
3	23	43	63	83		3	23	43	63	83
5	25	45	65	85		6	26	46	66	86
7	27	47	67	87		7	27	47	67	87
9	29	49	69	89		10	30	50	70	90
11	31	51	71	91		11	31	51	71	91
13	33	53	73	93		14	34	54	74	94
15	35	55	75	95		15	35	55	75	95
17	37	57	77	97		18	38	58	78	98
19	39	59	79	99		19	39	59	79	99

	CARD 3						CARD 4		
4	22	44	62	79		8	26	44	62	88
5	23	45	63	84		9	27	45	63	89
6	28	46	68	85		10	28	46	72	90
7	29	47	69	86		11	29	47	73	91
12	30	52	70	87		12	30	56	74	92
13	31	53	71	92		13	31	57	75	93
14	36	54	76	93		14	40	58	76	94
15	37	55	77	94		15	41	59	77	95
20	38	60	78	95		24	42	60	78	
21	39	61		100		25	43	61	79	

	CARD 5						CARD 6		
16	26	52	62	88		32	42	52	62
17	27	53	63	89		33	43	53	63
18	28	54	80	90		34	44	54	96
19	29	55	81	91		35	45	55	97
20	30	56	82	92		36	46	56	98
21	31	57	83	93		37	47	57	99
22	48	58	84	94		38	48	58	100
23	49	59	85	95		39	49	59	
24	50	60	86			40	50	60	
25	51	61	87			41	51	61	

CARD 7

64	74	84	94
65	75	85	95
66	76	86	96
67	77	87	97
68	78	88	98
69	79	89	99
70	80	90	100
71	81	91	
72	82	92	
73	83	93	

19.

```
C       PRIME NUMBER GENERATOR (X**2 + X + 41)

        PRINT 100

00100 FORMAT(72H1THE POLYNOMIAL (X**2 +X +41) WILL GENERATE THE FOLLOWIN
     1G PRIME NUMBERS ,//,11H VALUE OF X ,10X,12HPRIME NUMBER )

        K=0

00200 KPRIME = K**2 + K + 41

        PRINT 300,K,KPRIME

00300 FORMAT(1H0,4X,I2,18X,I5)

        IF(K-40)500,500,400

00400 STOP

00500 K=K+1

        GO TO 200

        END
```

21.

```
C       PRIME NUMBER GENERATOR (X**2 + X + 17)

        PRINT 10

00010 FORMAT(74H1THE POLYNOMIAL (X**2 +X +17) WILL GENERATE THE FOLLOWIN
     1G 16 PRIME NUMBERS,//,11H VALUE OF X,10X,12HPRIME NUMBER )

        N=0

00020 IPRIME = N**2 + N + 17

        PRINT 30,N,IPRIME

00030 FORMAT(1H0,4X,I2,19X,I3)

        IF(N - 14)40,40,50

00040 N=N+1

        GO TO 20

00050 STOP

        END
```

23. The given 5 by 5 square of numbers was not a magic square because row 5, main diagonal 2 and column 1 do not add up to the magic number 65.

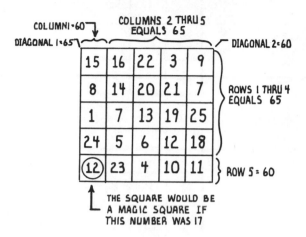

25. 102

27.

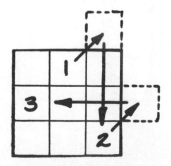

29. (1) Start with the number 1 in the center of row 1. Move diagonally to the right and if you leave the box, go to the bottom of the column in which you wanted to place the number. Put the number 2 in this box. Now again move diagonally to the right and if you leave the box, go to the left end of the row in which you wanted to place the number. Put the number 3 in this location. This completes a group of 3 numbers.

(2) Since this is a 3 by 3 square it is necessary to move down one

box to generate the next of 3 numbers. Thus 4, 5 and 6 are the next set of 3 numbers and are placed as follows;

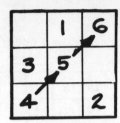

(3) The 7 is placed under the 6 because 6 ended a set of 3 numbers. The 8 and 9 are placed using the above rules and the completed magic square is pictured below.

8	1	6
3	5	7
4	9	2

31.

128	4	8
1	16	256
32	64	2

33.

8	3	4
1	5	9
6	7	2

35.

2^7	2^2	2^3
2^0	2^4	2^8
2^5	2^6	2^1

37. The values for a, h, i, g and e are determined in the noted order:

$$a = 2^3 = 2 \times 2 \times 2 = 8$$
$$h = 7$$
$$i = 2$$
$$g = 15 - 7 - 2 = 6$$
$$e = \frac{6 + 7 + 2}{3} = 5$$

The calculated values (a,h,i,g and e) are first placed in the blank 3 by 3 square. The remaining numbers are determined by selecting numbers that will produce a sum of 15 in each column, row and main diagonal.

8	1	6
3	5	7
4	9	2

39. The numerical values for C, D and E are

$$C = 2^3 = 2 \times 2 \times 2 = 8$$
$$D = 7$$
$$E = \frac{27}{3} - \frac{9}{3} - \frac{3}{3} = 9 - 3 - 1 = 5$$

The numerical values for C, D and E are first placed in the blank square. The remaining numbers are determined by selecting numbers that will produce a sum of 15 in each column, row and main diagonal.

6	7	2
1	5	9
8	3	4

Bibliography

BOOKS

Game Playing With Computers

Ahl, D. H., *BASIC Computer Games*, Maynard, Mass.: Digital Equipment Corp., 1973.

————, *Understanding Mathematics and Logic Using BASIC Computer Games*, Maynard, Mass.

Boehm, G. A. W., *The New World of Math*, New York: The Dial Press, 1959.

Davis, P. J., *The Lore of Large Numbers*, New York: Random House, Inc., 1961.

Desmonde, W. H., *Computers and Their Uses*, Englewood Cliffs, N. J.: Prentice-Hall, Inc., 1964.

Dorf, R. A., *Introduction to Computers and Computer Science*, San Francisco: Boyd & Fraser Publishing Co., 1972.

Dwyer, T. A. and M. S. Kaufman, *A Guided Tour of Computer Programming in BASIC*, Boston, Mass.: Houghton Mifflin Co., 1973.

Epstein, R. A., *The Theory of Gambling and Statistical Logic*, New York: Academic Press, 1967.

Feigenbaum, E. A. and J. Feldman, *Computers and Thought*, New York: McGraw-Hill Book Company, Inc., 1963.

Fink, D. G., *Computers and the Human Mind*, Garden City, New York: Doubleday & Co., 1966.

Gross, J. L. and W. S. Brainer, *Fundamental Programming Concepts*, New York: E. P. Dutton & Co.

Gruenberger, F. J. and D. D. McCracken, *Introduction to Electronic Computers,* New York: John Wiley & Sons, Inc., 1961.

Gruenberger, F. J. and G. Jaffray, *Problems For Computer Solution,* New York: John Wiley & Sons, Inc., 1965.

Halacy, D. S., *Computers: The Machines We Think With,* New York: Harper & Row, Publishers, 1962.

Harrison, J. O., *Computer-Aided Information Systems For Gaming,* U. S. Department of Commerce, AD 623091, September 1964.

Hawkes, N., *The Computer Revolution,* New York: E. P. Dutton & Co., 1972.

Jackson, P. C. Jr., *Introduction to Artificial Intelligence,* New York: Petrocelli Books, 1974.

Kochenburger, R. J. and C. J. Turcio, *Computers in Modern Society,* Santa Barbara, Calif.: Hamilton Publishing Co., 1974.

Litwin, S., R. McNaughton and R. L. Wexelblat, *General Switching Theory,* U. S. Department of Commerce, Office of Technical Services, PB 171555, January 1961.

McCracken, D. D., *A Guide To Fortran IV Programming,* New York: John Wiley & Sons, Inc., 1965.

———, *Digital Computer Programming,* New York: John Wiley & Sons, Inc., 1957.

Morrison, P. and E. Morrison, *Charles Babbage and His Calculating Engine,* New York: Dover Publications, Inc., 1961.

Newman, J. R., *The World of Mathematics* (Volume 4), New York: Simon and Schuster, Inc., 1956.

Organick, E. I., *A Fortran Primer,* Reading, Massachusetts: Addison-Wesley Publishing Company, Inc., 1963.

Pfeiffer, J., *The Thinking Machine,* Philadelphia: J. B. Lippincott Company, 1962.

Sage, E. R., *Fun and Games with the Computer,* Newburyport, Mass.: Entelek Inc., 1975.

Samuel, A. L., "Programming Computers to Play Games," *Advances in Computers* (Volume 1), New York: Academic Press, 1960.

Sherman, P. M., *Programming and Coding Digital Computers,* New York: John Wiley & Sons, Inc., 1963.

Snyder, G. S., *Let's Talk About Computers*, Middle Village, N. Y.: Jonathan David Publishers, 1973.

Spencer, D. D., *A Guide to BASIC Programming*, Sec. Ed., Reading, Mass.: Addison Wesley Publishing Co., 1975.

——, *Computers in Society*, Rochelle Park, N.J.: Hayden Book Co., 1974.

——, *Game Playing with BASIC*, Rochelle Park, N.J.: Hayden Book Co., 1976.

Trakhtenbrot, B. A., *Algorithms and Automatic Computing Machines*, Boston: D. C. Heath and Company, 1963.

Wilson, A. N., *The Casino Gambler's Guide*, New York: Harper & Row, Publishers, 1965.

Games

Adler, I., *Magic House of Numbers*, New York: The John Day Company, Inc., 1957.

Arnold, P., *The Book of Gambling*, London: The Hamlyn Publishing Group, Ltd., 1974.

Ball, W. W. Rouse, *Mathematical Recreations & Essays*, New York: The Macmillan Company, 1962.

Bell, R. C., *Board and Table Games*, London: Oxford University Press, 1960.

Bowers, H. and J. E. Bowers, *Arithmetical Excursions*, New York: Dover Publications, Inc., 1961.

Collver, D. I., *Scientific Blackjack & Complete Casino Guide*, New York: Arco Publishing Co., 1966.

Culin, S., *Games of the Orient*, Rutland, Vermont: Charles E. Tuttle Company, 1960.

Dana, J., *Black Jack—How to Win the Las Vegas Way*, Las Vegas, Nevada: Gambling International, 1965.

Domoryad, A. P., *Mathematical Games and Pastimes*, New York: The Macmillan Company, 1964.

Dudeney, H. E., *Amusements in Mathematics*, New York: Dover Publications, Inc., 1958.

————, *The Canterbury Puzzles*, New York: Dover Publications, Inc., 1958.

Eves, H., *An Introduction to the History of Mathematics*, New York: Rinehart and Company, Inc., 1953.

Frey, R. L., *According to Hoyle*, Greenwich, Conn.: Fawcett Publications, Inc., 1970.

Friedman, B., *Casino Games*, New York: Western Publishing Co., 1973.

Gardner, M., *Mathematical Puzzles and Diversions*, New York: Simon and Schuster, Inc., 1959.

————, *Mathematical Puzzles of Sam Loyd*, Volume 1, New York: Dover Publications, Inc., 1959.

————, *Mathematical Puzzles of Sam Loyd*, Volume 2, New York: Dover Publications, Inc., 1960.

————, *New Mathematical Diversions From Scientific American*, New York: Simon and Schuster, Inc., 1966.

Golomb, S. W., *Polyominoes*, New York: Charles Scribner's Sons, 1965.

Goodman, M., *Slots and Pinballs—How To Win The Las Vegas Way*, Las Vegas, Nevada: Gambling International, 1965.

Goren, E. H., *Goren's Hoyle Encyclopedia of Games*, New York: Hawthorn Books Inc., 1961.

Heath, R. V., *Mathe-E-Magic*, New York: Dover Publications, Inc., 1953.

Hervey, G. F., *The Complete Illustrated Book of Card Games*, Garden City, New York: Doubleday & Co., 1973.

Hollings, F., *The Beginner's Book of Chess*, Philadelphia: David McKay Company, 1930.

Hopper, M., *Win at Backgammon*, New York: Dover Publications, 1972.

Horowitz, A., *Chess For Beginners*, New York: Barnes & Noble, Inc., 1956.

Hunter, J. A. H., *Mathematical Diversions,* Princeton, N. J.: D. Van Nostrand Co., 1963.

Jacoby, O., *How to Figure the Odds,* Garden City, New York: Doubleday & Company, Inc., 1947.

Jarlson, G., *Roulette—How to Win the Las Vegas Way,* Las Vegas, Nevada: Gambling International, 1965.

Johnson, D. A. and W. H. Glenn, *Exploring Mathematics on Your Own,* Garden City, New York: Doubleday & Company, Inc., 1961.

Kishikawa, S., *Steppingstones to GO: A Game of Strategy,* Rutland, Vermont: Charles E. Tuttle Company, Inc., 1965.

Kraitchik, M., *Mathematical Recreations,* New York: Dover Publications, Inc., 1953.

Lemmel, M., *Gambling: Nevada Style,* Garden City, New York: Doubleday & Company, Inc., 1964.

Levinson, H. C., *The Science of Chance,* New York: Rinehart & Company, 1950.

McQuaid, C., *Gambler's Digest,* Chicago: Follett Publishing Co., 1971.

Menninger, K. W., *Mathematics In Your World,* New York: The Viking Press, 1962.

Meyer, J. S., *More Fun With Mathematics,* Greenwich, Connecticut: Fawcett Publishing Company, 1963.

Morehead, A. H., R. L. Frey and G. Mott-Smith, *The New Complete Hoyle,* Garden City, New York: Garden City Books, 1964.

Mott-Smith, G., *Mathematical Puzzles,* New York: Dover Publications, Inc., 1954.

Mulac, M. E., *The Game Book,* New York: Harper & Brothers Publishers, 1946.

Murray, H. J. R., *A History of Board-Games Other Than Chess,* London: Oxford University Press, 1952.

Newman, J. R., *The World of Mathematics,* Volume 4, New York: Simon and Schuster, Inc., 1956.

Niven, I., *Numbers: Rational and Irrational,* New York: Random House, Inc., 1961.

O'Beirne, T. H., *Puzzles and Paradoxes*, New York: Oxford University Press, 1965.

O'Neil-Dunne, P., *Roulette for the Millions*, Chicago: Henry Regnery Co., 1972.

Ore, O., *Number Theory and Its History*, New York: McGraw-Hill Book Company, Inc., 1948.

Parker, H. W., Jr., *Bingo-Keno—How to Win the Las Vegas Way*, Las Vegas, Nevada: Gambling International, 1965.

Pedoe, D., *The Gentle Art of Mathematics*, New York: The Macmillan Company, 1958.

Rademacher, H. and O. Toeplitz, *The Enjoyment of Mathematics*, Princeton, N. J.: Princeton University Press, 1957.

Radner, S. H., *The Key to Roulette, Blackjack, and One-Armed Bandits*, —— —: Ottenheimer Publishers, Inc., 1963.

Revere, L., *Playing Blackjack As a Business*, Secaucus, N. J.: Lyle Stuart, Inc., 1973.

Rice, T., *Mathematical Games and Puzzles*, New York: St. Martin's Press, 1973.

Scarne, J., *Scarne's Complete Guide to Gambling*, New York: Simon and Schuster, 1961.

———, *Scarne's Encyclopedia of Games*, New York: Harper & Row, 1973.

Scott, J. and L. Scott, *Fun and Games with Cards*, New York: Ace Books, 1973.

Smith, A., *The Game of GO*, Rutland, Vermont: Charles E. Tuttle Company, Inc., 1956.

Smith, D. E., *History of Mathematics*, Volume 2, New York: Dover Publications, Inc., 1958.

Smith, H. S., Sr., *I Want To Quit Winners*, Englewood Cliffs, N. J.: Prentice-Hall, Inc., 1961.

Squire, N., *How to Win at Roulette*, Las Vegas, Nevada: Gambler's Book Club, 1972.

Stewart, B. M., *Theory of Numbers,* New York: The Macmillan Company, 1952.

Thorp, E. O., *Beat the Dealer,* New York: Random House, 1962.

Wilson, A. N., *The Casino Gambler's Guide,* New York: Harper & Row, Publishers, 1965.

Woon, B., *Gambling in Nevada,* Reno, Nevada: Bonanza Publishing Company, 1953.

Wykes, A., *The Complete Illustrated Guide to Gambling,* New York: Doubleday & Company, Inc., 1964.

Yaglom, A. M. and I. M. Yaglom, *Challenging Mathematical Problems with Elementary Solutions,* San Francisco: Holden-Day, Inc., 1964.

Magic Squares

Andrews, W. S., *Magic Squares and Cubes,* New York: Dover Publications, Inc., 1960.

Ball, W. W. R. and H. S. M. Coxeter, *Mathematical Recreations and Essays,* 11th edition, New York: The Macmillan Company, 1938.

Bowers, H. and J. E. Bowers, *Arithmetical Excursions,* New York: Dover Publishing, Inc., 1961.

Cajori, F., *A History of Mathematics,* New York: The Macmillan Company, 1919.

Davies, C. and W. G. Peck, *Mathematical Dictionary and Cyclopedia of Mathematical Science,* New York: A. S. Barnes and Company, 1855.

Domoryad, A. P., *Mathematical Games and Pastimes,* New York: The Macmillan Company, 1964.

Dudeney, H. E., *Amusements In Mathematics,* New York: Dover Publishing, Inc., 1958.

Encyclopedia Britannica, Volume 14, Encyclopedia Britannica, Inc., 1962, Pages 625-27.

Eves, H., *An Introduction to the History of Mathematics,* New York: Rinehart and Company, Inc., 1953.

Falkner, E., *Games Ancient and Oriental and How to Play Them,* New York: Dover Publishing, Inc., 1961.

Fults, J. L., *Magic Squares*, La Salle, Illinois: Open Court, 1974.

Gardner, M., *The 2nd Scientific American Book of Mathematical Puzzles & Diversions*, New York: Simon and Schuster, 1961.

Heath, R. V., *Math E Magic*, New York: Dover Publications, Inc., 1953.

Hogben, L., *Mathematics in the Making*, Garden City, New York: Doubleday & Company, Inc., 1960.

Kraitchik, M., *Mathematical Recreations*, New York: Dover Publications, Inc., 1953.

Merrill, H. A., *Mathematical Excursions*, New York: Dover Publications, Inc., 1933.

Meyer, J. S., *Arithmetricks*, New York: Scholastic Magazines, Inc., 1965.

————, *Fun With Mathematics*, Greenwich, Connecticut: Fawcett Publishing, Inc., 1963.

————, *More Fun with Mathematics*, Greenwich, Connecticut: Fawcett Publishing, Inc., 1963.

Sanford, V., *A Short History of Mathematics*, Boston: Houghton Mifflin Company, 1930.

Simon, W., *Mathematical Magic*, New York: Charles Scribner's Sons, 1964.

Smith, D. E., *History of Mathematics*, Volume 1, New York: Dover Publications, Inc., 1958.

————, *History of Mathematics*, Volume 2, New York: Dover Publications, Inc., 1958.

————, *Number Stories of Long Ago*, Washington, D. C.: The National Council of Teachers of Mathematics, 1962.

White, W. F., *A Scrap-Book of Elementary Mathematics*, Chicago: The Open Court Publishing Company, 1908.

Programming Languages

Coan, J. S., *Basic BASIC: An Introduction to Computer Programming in BASIC Language*, Rochelle Park, N. J.: Hayden Book Co., 1970.

Dawson, C. B. and T. C. Wool,. *From Bits to Ifs,* New York: Harper & Row, 1971.

Dwyer, T. A. and M. S. Kaufman, *A Guided Tour of Computer Programming in BASIC,* Boston: Houghton Mifflin Co., 1973.

Gateley, W. Y., and G. G. Bitter, *BASIC for Beginners,* New York: McGraw-Hill Book Co., 1970.

Hare, V. C., *BASIC Programming,* New York: Harcourt, Brace & World, 1970.

Maniotes, J., H. Higley, and J. N. Haag, *Beginning FORTRAN,* Rochelle Park, N. J.: Hayden Book Co., 1969.

Pavlovich, J. P. and T. E. Tahan, *Computer Programming in BASIC,* San Francisco: Holden Day, 1971.

Spencer, D. D., *A Guide to BASIC Programming,* Reading, Mass.: Addison-Wesley Publishing Co., 1975.

Wilf, H. S., *Programming for a Digital Computer in the FORTRAN Language,* Reading, Mass.: Addison-Wesley Publishing Co., 1969.